THE

CAMBRIAN COAST

RAILWAY

Morfa Mawddach, 7th June 1963 *Photo: B. J. Ashworth*

COMPILED and WRITTEN by

W. G. REAR & M. F. WILLIAMS

ISBN 1-870119-03-7

Designed and edited by Gregory K. Fox
Printed by Netherwood Dalton & Co. Ltd., Huddersfield
Published by Foxline Publishing
32, Urwick Road, Romiley, Stockport.

Penhelig Halt, 19th June 1962. Photo: B. J. Ashworth

The Cambrian Coast Railway – Dovey Junction to Pwllheli

In the eighteen thirties, Brunel had planned an Irish mail route through Ludlow, Newtown and Barmouth to Porth Dinllaen. There was also a proposal by the Great North & South Wales & Worcester Railway to link Carmarthen with Porth Dinllaen via Cardigan, Aberystwyth, Machynlleth & Dolgellau. The GWR also suggested a Worcester to Porth Dinllaen route to complete a trio of schemes which came to nought. In addition bills were proposed from the West Midland, Shrewsbury & Coast of Wales Railway, and Portmadoc & Porth Dinllaen Railway, both of which failed in Parliament.

The Aberystwyth & Welch (sic) Coast Railway however was successful in gaining Assent by Act of 22nd July 1861 for a line from Aberystwyth through Towyn, Barmouth and Portmadoc to Pwllheli, with a connection south of the Dovey to the Newtown & Machynlleth line at the latter. The Engineer to the line was Benjamin Piercy with Savin as the contractor.

The section from Machynlleth to Aberystwyth was built first with the portion from Machynlleth to Borth opening on 1st July 1863 and the remaining 8¾ miles opening to passenger traffic on 23rd June 1864. The line north of the Dovey estuary was opened in sections between Aberdovey (harbour) and Llwyngwril with an intermediate station at Towyn. Savin the contractor had ferried locomotives, subsequently absorbed into Cambrian Railways' stock, to a temporary siding at Ynyslas. After delivery of coaching stock to Aberdovey the line opened to the public on 24th October 1864 with three trains each way, daily between the two points.

The section across the Dovey estuary between Aberdovey harbour and the Machynlleth to Aberystwyth line at Ynyslas was subsequently considered too difficult and so a deviation was proposed which followed the northern coastline of the river, taking the line through the back of Aberdovey by way of four short tunnels, not to mention some heavy shore line earth works and stone retaining walls.

This legislation incorporated reciprocal running powers over the Wnion and Dee Valley lines to enable users of the lines to reach Aberystwyth, although the connection with the Machynlleth to Aberystwyth line was such that it was not possible to run through trains from Aberystwyth to points along the Cambrian Coast without reversal. The A & WC Railway Bill received Royal Assent on 5th July 1865, the same day on which the Cambrian and Coast Railways (Amalgamation) Bill incorporated the A & WC Railway into the Cambrian system. The Dovey was finally crossed to the satisfaction of the Inspecting Officer and services commenced on 14th August 1867.

Work proceeded apace north of Barmouth and subsequently an inspection took place of the 33¼ miles between the junction and Pwllheli. The bridge over the Mawddach was officially passed for passenger traffic in September 1867. The Carnarvonshire Railway started running trains to Afonwen on 2nd September of the same year and services were extended east to Penrhyndeudraeth until the Cambrian commenced running of their own trains on 10th October 1867.

From time to time, various proposals to extend the Cambrian line to Porth Dinllaen surfaced, the final attempt being made in 1913 when the Cambrian obtained powers for an extension of their line. World War I intervened and the situation afterwards did not justify the proposal to materialise. An extension at Pwllheli was put into operation on 19th July 1909 when a new terminus was opened some forty chains west of the original terminus.

Llangelynin, 14th June 1965. Skirting the rocky shoreline south of Llangelynin is BR Standard Class 4MT locomotive No. 75002 with the southbound or Up portion of the 'Cambrian Coast Express' on the first day of the 1965 summer timetable. On weekdays the train departed from Pwllheli at 8.20a.m. to link up at Dovey Junction (arr 10.30 a.m.) with the Aberystwyth portion. At the time it was still possible for the princely sum of two shillings, to reserve seats in advance at many of the coast stations. After reversal at Shrewsbury, the train travelled via Wolverhampton Low Level, Birmingham Snow Hill, Leamington Spa General and Banbury before arrival at Paddington at 4p.m. *Photo: B. J. Ashworth*

Proposals were made for shipping services from the Cambrian coast and in 1883 powers were obtained to extend Aberdovey harbour for the shipping of slate. The Aberdovey & Waterford Steam Shipping Company put on two passenger steamers, the *Cambria* and *Liverpool*. In 1889 the Cambrian was authorised to operate steamers between Aberdovey, Aberystwyth, Portmadoc and Pwllheli and Irish ports but all ventures were unsuccessful and the services were short lived.

Geographically the line followed the coast for almost its entire length, and rarely ventured far inland. Nevertheless it was not without severe gradients, and although tunnelling was necessary in several places, none were extensive. Penhelig Halt, located between two tunnels within a half mile presented problems to footplate crews working long trains where piloting of the train engine was necessary, the confined bores combining with the slow start pulling away from rest ensuring that enginemen gained the full benefit of inhaling smoke! One didn't hang about in these conditions, and as regular double heading of some trains was necessary, train crews ensured that under these conditions, everyone pulled their weight. There was, and still is, a very effective method of levelling out old scores in these conditions, especially if the offended party is on the leading engine, and the offender is on the train locomotive.

South of Barmouth, the line climbed to its highest point at Friog – between Fairborne and Llwyngwril – scene of the two accidents in 1883 and almost identically again in 1933, when trains left the rails and plunged to the shore below with fatal results. Somewhat belatedly, an avalanche shelter was built over the line to offer protection. Further north where the line follows the estuary inland from Llandecwyn to Penrhyndeudraeth, it climbs beyond Penrhyn up a 1 in 102 gradient to the summit, falling away again to Minffordd at 1 in 50. This necessitated assisting most freight trains between the two points, and in fact Portmadoc shed had pilot engine duties specifically for this purpose. It was the practice to assist trains from Portmadoc on the southbound journey, detaching the locomotive at Penrhyndeudraeth. Freight traffic was sufficiently heavy to justify two turns for this purpose until the end of steam. Also, shunting turns were required to assist if necessary, and this was indicated in the service timetables.

The route was classified as 'yellow' until bridge strengthening was effected, enabling 'blue' route locomotives to work over the lines, but with one or two exceptions, such as sidings and where curves were considered too tight for such motive power.

With the coming of the railway, many small settlements developed, and tourism provided welcome employment in its own right. The railway was for several decades the main employer and staff were respected members of the community, particularly the stationmaster, although in all probability his wages did not reflect this status. It was the 'rationalization and modernization' plans of Dr. Beeching that eliminated all but a handful of staff and the stationmaster disappeared from the community together with porters, signalmen and traincrews. Today, Barmouth is one of very few stations which retain any staff other than signalmen, at Dovey Junction, Harlech and Portmadoc. Traincrews formerly based at Portmadoc have also gone and Pwllheli, which had 17 sets of drivers and firemen in December 1964, has seen its compliment reduced to four drivers. In 1938 there were fourteen rostered duties at Portmadoc, Machynlleth having thirty six of which at least ten worked along the coast line. Today there are but sixteen drivers remaining.

Acknowledgements

A work of this nature depends largely on the assistance and efforts of many people. We have been extremely fortunate to have access to the work of many excellent photographers who have provided material freely from their collections, often at some inconvenience to themselves. To the following we extend our sincere thanks:- B. J. Ashworth, P. E. Baughan, J. M. Bentley, J. R. Besley, C. L. Caddy, who also made available his collection taken by the late J. W. T. (Joe) House, the late Derek Cross, J. J. Davis, Alan Donaldson, Michael Hale, E. N. Kneale, R. W. Hinton, L. Hobdey, Norman Jones, Lens of Sutton, the late T. Lewis, L. Marshall, M. Mensing, J. H. (Joe) Moss, the late J. K. Nelson, Mrs. V. Palmer, the late G. H. Platt, H. B. Priestley, R. C. Riley, R. F. Roberts, British Rail and M. S. Welch.

Others have contributed through their knowledge and recollections and thanks must be proffered to the staff of British Rail on the Cambrian Coast line both past and present, especially those at Pwllheli, Barmouth and Machynlleth. We would particularly mention the assistance given by the late Mr. Starling, one time driver at Portmadoc. A mention must also be made of the help given by John Kimberley, friend and fellow enthusiast on the Llangollen and Bala Lake railways, for his assistance in proof reading and research. To Jacqueline and Greg Fox for providing maps, drawings and many other forms of practical assistance, including the faith in us to publish this work.

Finally, it must be placed on record the debt owed to our respective families for their tolerance and understanding during the preparation of this book. Our wives, Norma and Susan, have provided support and encouragement without complaint during the frequent and sometimes long periods of absence.

W. G. Rear, M. F. Williams, Johnstown and Rhos, Nr. Wrexham

Footnote: The place names in this book are essentially railway orientated and spelt in such a way as to be contemporary with the date of the photograph, map or timetable, etc.

Above: Reproduction of the cover for the Great Western Railway publication of 1936, "Rambles around the Cambrian Coast", maps from which are included in this book adjacent to the respective locations.

Below: Reproduction of the cover for the Great Western Railway Service Time Tables, Whitchurch, Brecon, Pwllheli and Aberystwyth area, July 4th to September. Extracts of coast line train services are detailed on the next few pages.

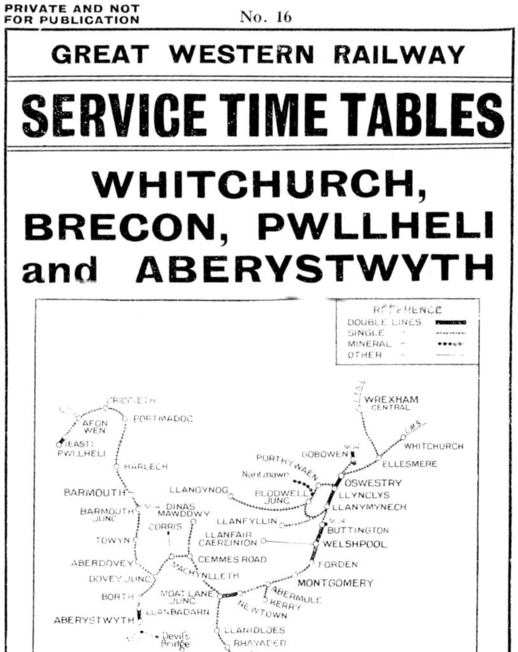

GRADIENT PROFILE - PWLLHELI to DOVEY JUNCTION

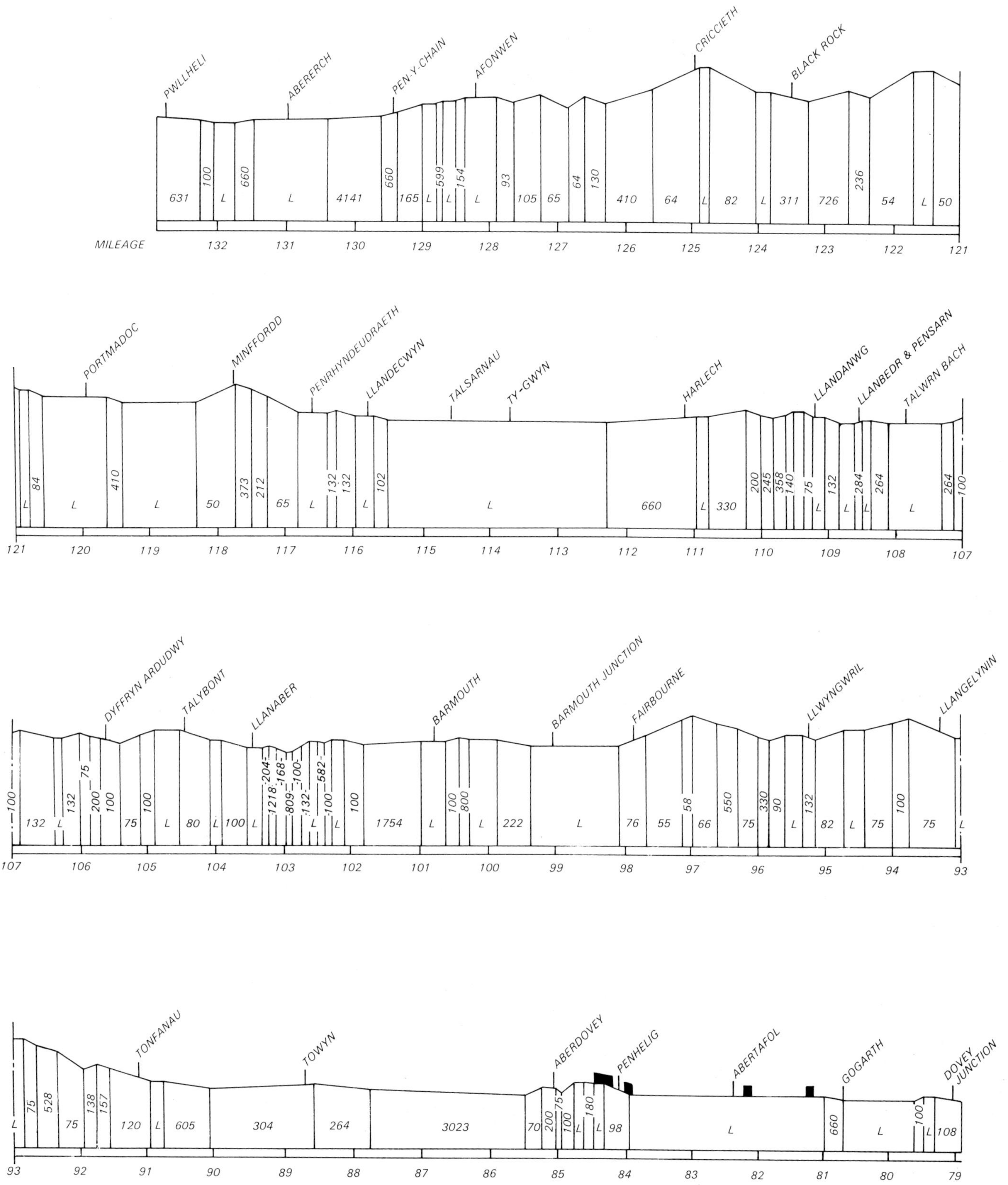

GKF 88

Train Services

From Dovey Junction to Pwllheli the route was and is single line, worked by Electric Token system with passing loops, although some sections were double line and worked by the Absolute Block system. The sections between Barmouth South and Barmouth North, Pwllheli East and Pwllheli (station) were so worked until 1947. Similarly operated was the section from Afonwen to Penychain which was doubled to coincide with the opening of Butlin's Holiday Camp at Penychain, enabling quick disposal of trains from the two platforms.

Crossing stations were located at Dovey Junction, Aberdovey, Towyn, Llwyngwril, Barmouth Junction, Dyffryn Ardudwy, Llanbedr & Pensarn, Harlech Penrhyndeudraeth, Portmadoc and Criccieth. These crossing points were reduced in number with the elimination of steam and freight services. Today, crossing points are located at Dovey Junction, Towyn, Barmouth, Harlech and Portmadoc. Portmadoc to Pwllheli being no more than an extended siding.

Journey times were not fast, due to a maximum speed limit of 55mph over the whole route. Certain trains covered the 53 miles 2 chains faster than others, bypassing the smaller halts. Through trains were few in number until after the war although the Cambrian Railways were enthusiastic exponents of through carriage workings. In pre-war peak summer times, through trains from Wrexham, Ruabon or Birkenhead to Pwllheli were confined to a couple of trips daily, with the corresponding workings in the Up direction. Through workings to or from points other than those mentioned above were confined to unbalanced Saturdays workings to Birmingham Snow Hill, from Pwllheli and to Paddington, from Barmouth. Examination of the 1938 Service Timetable will clarify the picture. It should be noted that freight workings generally only worked part of the route. Portmadoc shed had an interesting working, providing assistance to the passenger turn from Pwllheli to Machynlleth. The engine was attached at Portmadoc as pilot and worked to Towyn, where it was detached, subsequently taking the 8.05a.m. goods to Pwllheli, the crew working this duty being relieved at Portmadoc at 11.15a.m. by a second pair who continued to Pwllheli East.

Prior to Nationalization, although there were running powers and agreements between the LMS and GWR, there were no regular week-day workings by LMS crews between Afonwen and Pwllheli. However, on Sundays in the high summer season two trains from Bangor ran round the stock at Afonwen and worked to Pwllheli, a turn covered by LMS crews. The GWR never invoked their rights although there is some indication that on one or two occasions in 1938 and 1939, excursion trains which started at Chwilog were run by Portmadoc men as far as Machynlleth, at which point they were handed over to other crews with through trains to Shrewsbury. The events and dates are not recorded, but according to railwaymen

Down Trains. MACHYNLLETH, BARMOUTH, PORTMADOC and PWLLHELI. Week Days.

No. 16

Miles from M'chynlleth M.	C.	Miles from Whitchurch and M.P. Mileage M.	C.	STATIONS.	Ruling Gradient 1 in	Allow for Stop (Mins.)	Allow for Start (Mins.)	"C" Head Lamp Trains (Mins.)	"D" Head Lamp Trains (Mins.)	Express Trains (Mins.)	Ordinary Trains (Mins.)
..	..	75	11	cMACHYNLLETH	132 R.		1				
3	72	79	3	cDovey Junction	132 F.	1	1				9
5	43	80	54	Gogarth Halt							
7	18	82	29	Abertafol Halt							
8	77	84	8	Penhelig Halt							
9	71	85	2	cAberdovey	98 R.	1	1				17
13	46	88	57	cTowyn	70 F.	1	1			33	10
15	67	90	78	Tonfanau Quarry Siding							
15	78	91	9	dTonfanau	120 R.						
18	13	93	24	Llangelynin Halt							
20	8	95	19	cLlwyngwril	75 RF	1	1				19
21	43½	96	54½	Stop Board							
22	61	97	72	Fairbourne	66 R. 55 F.	1	1				8
23	53	98	64	Dolgelley Branch Loop							
23	69	99	0	cBarmouth Junction	76 F.	1	1			27	2
25	39	100	50	cBARMOUTH	222 R.	1	1			5	5
27	3	102	14	Llanaber Halt							
29	26	104	37	Talybont Halt							
30	39	105	50	cDyffryn-on-Sea	75 R.	1	1				15
32	57	107	68	Talwrn Bach Halt							
33	34	108	45	cLlanbedr and Pensarn	75 F.	1	1				8
34	6	109	17	Llandanwg Halt							
36	2	111	13	cHarlech	75 R.	1	1				8
38	44	113	55	Tygwyn Halt							
39	32	114	43	Talsarnau	660 F.	1	1				9
40	52	115	63	Llandecwyn Halt							
41	36	116	47	cPenrhyndeudraeth	102 R.	1	1				8
42	47	117	58	dMinffordd	50F, 65R	1	1				7
43	17	118	28	Granite Quarry Siding							
44	66	119	77	cPORTMADOC	50 F.	1	1			50	5
46	45	121	56	Wern Siding							
48	27	123	38	Black Rock Halt							
49	63	124	74	cCriccieth	50 R. 54 F.	1	1				11
53	6	128	17	cAfon Wen	64 F.	1	1			19	6
54	22	129	33	Penychain Halt							
55	66	130	77	Abererch							
57	4	132	15	cbPwllheli (East) (Double Line)	100 R.	1				8	8
57	54	132	65	bPWLLHELI (Double Line)	631 R.						

Time Allowances for Freight Trains—see page 2. Point-to-Point Allowances.

STATIONS.	F Goods arr.	F Goods dep.	B Pass. and Mail dep.	K Pilot arr. / (below) B Passenger arr.	K Pilot dep. / (below) B Passenger dep.	F 4.3 a.m. Goods and Mail ex Ruabon arr.	F dep.	K Goods dep.	B Passenger and Mail arr.	B Passenger and Mail dep.	B Passenger 7.0 a.m. ex Dolgelley arr. / (below) B Passenger arr.	B dep. / (below) B Passenger dep.	C Stores SO arr.	C Stores SO dep.
	a.m.	a.m.	a.m.	a.m.	a.m.	a.m.	a.m.	a.m.	a.m.	a.m.	a.m.	a.m.	a.m.	a.m.
MACHYNLLETH	—	5 15							T	6 10			—	6 30
Dovey Junction	5 25								6 16				6 37	
Gogarth Halt	—	—							—	—			—	—
Abertafol Halt	—	—							—	—			—	—
Penhelig Halt	—	—							—	—			—	—
Aberdovey	—	—							—	6 32			6 51	6 53
Towyn	5 50	—		—	6 0				—	6 39			6 59	7 3
Tonfanau Quarry Siding				6 7	7 18				—	—			—	—
Tonfanau									—	‡			—	7 8
Llangelynin Halt									—	—			—	—
Llwyngwril									—	6 52			7 15	7 17
Stop Board									—	—			—	—
Fairbourne									—	7 0			7 24	7 26
Dolgelley Branch Loop									—	—			—	—
Barmouth Junction						6 50			7 2½	7 4	7 21	7 22	7 29	X7 46
BARMOUTH						6 56	—		7 9	X7 40	7 27	X —	7 51	—
Llanaber Halt	Engine to work pilot trip to Tonfanau and back, afterwards shunt at Towyn until 7.55 a.m.								—	—			Will run July 16th to Sept. 10th, inclusive.	
Talybont Halt									—	—				
Dyffryn-on-Sea									—	7 48½				
Talwrn Bach Halt									—	—				
Llanbedr and Pensarn									—	7 54½				
Llandanwg Halt									—	—				
Harlech									—	8 2				
Tygwyn Halt									—	8 7				
Talsarnau									—	8 10				
Llandecwyn Halt									—	8 14½				
Penrhyndeudraeth									—	8 18½				
Minffordd									—	8 22				
Granite Quarry Siding									—	—				
PORTMADOC				—	6X40				8 26½	X8 30	—	9 0		
Wern Siding					—					—		—		
Black Rock Halt				—	—					8 37½	—	—		
Criccieth				—	6 49				8 40½	8 41	9 8	9 9		
Afon Wen			6T22	6 55	X6 57			7 15	8 46½	8 47	9 14½	X9 18		
Penychain Halt			—	—	—			—	—	8 50	—	—		
Abererch			—	—	7 2½			CR	—	8 53½	—	9 23½		
Pwllheli (East)			—	—	—			7 30	—	—	—	—		
PWLLHELI			6 32	7 8	—				8 59	—	9 29	—		

‡—Stops to set down on notice to Guard

? Penychain to Afonwen now double track — [illegible] Butlins Camp

Down Trains. MACHYNLLETH, BARMOUTH, PORTMADOC and PWLLHELI. Week Days.

No. 16

STATIONS.	G L.E. SO dep.	K Goods arr.	K Goods dep.	G L.E. SO C dep.	K 4.45 a.m. Ruabon Goods arr. / (below) B Passenger arr.	K dep. / (below) B Passenger dep.	F Goods SX arr.	F Goods SX dep.	F Goods SO arr.	F Goods SO dep.	G L.E. SO Z dep.	B Auto ex Dolgelley dep.	B Passenger arr.	B Passenger dep.	G L.E. RR dep.	B Passenger 7.33 a.m. ex Wrexham arr.	B dep.	K Pilot dep.	B Auto 10.30 a.m. ex Dolgelley SX arr.	B Auto SX dep.	A Passenger ex Aberystwyth WO arr.	A WO dep.
	a.m.	a.m.	a.m.	a.m.	a.m.	a.m.	a.m.	a.m.	a.m.	a.m.	a.m.	a.m.	a.m.	a.m.	a.m.	a.m.	a.m.	a.m.	a.m.	a.m.	a.m.	a.m.
MACHYNLLETH	7‖ 0	—	7 20										—	8 15							Y	—
Dovey Junction	7 7	7 30											8X21	T8 31							10 2	10 8
Gogarth Halt		—	—										—	8 35½							—	—
Abertafol Halt		—	—										—	8 41½							—	—
Penhelig Halt		—	—										—	8 46							—	—
Aberdovey	7 22	7 50X	●9 18										—	X8 49							10 22	X1024
Towyn	7X30	9 30	—				—	X8 5	—	8. 5			—	8X56	9‖11						—	10 31
Tonfanau Quarry Siding	—						—	—	—	—			—	—	—						—	—
Tonfanau	—						—	—	—	—			—	9 1	—						—	—
Llangelynin Halt	—						—	—	—	—			—	9 5½	—						—	—
Llwyngwril	§7X47						8C26	X8T 33	—	—			—	X9 10	—						—	—
Stop Board	—						P	—	—	P			—	—	—						—	—
Fairbourne	—						—	—	—	—			—	9 17½	—						—	—
Dolgelley Branch Loop	—						—	—	—	—			—	—	—						—	—
Barmouth Junction	8‖X0			8‖30	8 33	U8 48	8 47	●9 0	8 36	9 0	9‖33	9 11	9 20	X9 21	9X‖40	10 4	10 5		10 52	10X57	10 51	X11 3
BARMOUTH	8‖ 5			8‖36	8 55	—	9 7X	●9 55	9 7X	●9 50	9‖X39	9X16	9 26	X9 29		10X10	10¶23		11 2	—	11 8	11 10
Llanaber Halt	Whenever possible will work attached to 6.30 a.m. Stores Train. Will not run after September 3rd.						—	—	—	—			—	9 33	Does not run on Saturdays nor when Towyn Schools are closed.	—	10¶27				—	—
Talybont Halt							—	—	—	—			—	9 37½		—	10¶31				—	—
Dyffryn-on-Sea							—	—	10C7X	10S12			9 40½	X9 41		1034½	X1037				11	X12
Talwrn Bach Halt							—	—	—	—			—	9 46		—	10 42				—	—
Llanbedr and Pensarn							10V20	X10 32	10 22	X1032			—	9 49		—	10 45				—	—
Llandanwg Halt							—	—	—	—			—	9 51		—	1047½				—	—
Harlech							—	—	—	—			9 56½	X9 57		10 53	—				11 30	X1133
Tygwyn Halt							—	—	—	—			—	10 2							—	—
Talsarnau							—	—	—	—			—	10 5							—	—
Llandecwyn Halt							—	—	—	—			—	—				N			—	—
Penrhyndeudraeth							—	—	10 51	X1120			1010X	1013½							11 47	1147½
Minffordd							—	—	—	—			—	10 18							—	—
Granite Quarry Sidings							—	—	—	—			—	—				11 52			—	—
PORTMADOC					—	10 5	11 5	X11 30	11 34	X1238			1022½	10 30				12 0			11 52	—
Wern Siding					—	—	—	—	—	—			—	—							Forms 11.55 a.m. Portmadoc to Pwllheli.	
Black Rock Halt					—	1012½	—	—	—	—			—	—								
Criccieth					10 16	10 18	11 43	11 57	12 51	X1 17			1038½	X1040½								
Afon Wen					1023½	10X28	12 5X	●12 35	1 25	●2X 0			10 46	X1051								
Penychain Halt					—	10 31	—	—	—	—			—	—								
Abererch					—	1034½	—	—	—	—			—	1056½								
Pwllheli (East) (Double Line)					—	—	12 50	—	2 15	—			—	—								
PWLLHELI (Double Line)					10 40	—							11 2	—								

C—To work 9.55 a.m. Barmouth to Chester.
N—On the Wednesdays and Saturdays when this schedule is impracticable, Station Master, to advise the Quarry Manager and arrange a special pilot trip later in the day.
U—Engine to turn before proceeding to Barmouth.
V—Calls to detach cement traffic ex Aberdovey or grain traffic ex Birkenhead, otherwise for token only.
Y—Runs on Wednesdays, July 20th to September 7th, inclusive. §—Arrives Llwyngwril, 7.44 a.m. ¶—Advertised 3 minutes earlier.
Z—To work 10.50 a.m. Barmouth to Ruabon.

Down Trains. MACHYNLLETH, BARMOUTH, PORTMADOC and PWLLHELI. Week Days.

No. 16

STATIONS.	B 11.2 a.m. Auto ex Dolgelley SO arr. / (below) B Passenger arr.	B SO dep. / (below) B Passenger dep.	B Passenger SX arr.	B Passenger SX dep.	B Passenger SO arr. / (below) B Passenger arr.	B Passenger SO dep. / (below) B Passenger dep.	B 8.13 a.m. Passenger ex Birkenhead arr.	B dep.	K Goods arr.	K Goods dep.	G L.E. SO Y dep.	G L.E. N dep.	B Pass. SO dep.	K Goods arr.	K Goods dep.	B 1.43 p.m. Dolgelley Auto arr.	B dep.	B Passenger SO arr.	B Passenger SO dep.	B Passenger arr.	B Passenger dep.
	a.m.	a.m.	a.m.	a.m.	a.m.	a.m.	a.m.	a.m.	p.m.	p.m.	p.m.	p.m.	p.m.	p.m.	p.m.	p.m.	p.m.	p.m.	p.m.	p.m.	p.m.
MACHYNLLETH			—	10†15	—	10†15			Will not run on Saturdays, August 6th, 13th, 20th and 27th.				11.0 a.m. ex Ruabon.							—	12U40
Dovey Junction			10†22	T1055	10†22	11 5														1246X	12T49
Gogarth Halt			—	1059½	—	11 9½														—	1253½
Abertafol Halt			—	11 5½	—	1115½														—	1259½
Penhelig Halt			—	11 10	—	11 20														—	1 4
Aberdovey			— X	11 13	X	11 23														1 7	X1 10
Towyn			1119X	11 20	—	11 30								—	11 45					1 16	1 18
Tonfanau Quarry Siding				—	—	—								—	—					—	—
Tonfanau			—	11 25	—	11 35								—	—					—	1 23
Llangelynin Halt			—	11 29	—	11 39								—	—					—	1 28
Llwyngwril			—	11 34	—	11 44								12 6	X12 45					—	1 32
Stop Board			—	—	—	—								P	—					—	—
Fairbourne			—	11 42	—	1151½								1 0	1 15					—	1 40
Dolgelley Branch Loop			—		—	—								—	—					—	—
Barmouth Junction	11 25	11 26	11 45	11 46	1154X	11§55	11‡53	X1154			12‖49	12‖55	1X21	1 20X	●1 53	2 6	2 7			1 42½	1 43
BARMOUTH	11 31	—	11 51	—	12§ 0	—	11‡59	12X25	—	12 5	12‖54	1‖ 0	1X26	2 0	—	2 12	—			1 48	X2 15
Llanaber Halt							—	12 29	—	—								Will not run after September 10th.		—	2 19
Talybont Halt							—	12 34	—	—										—	2 23½
Dyffryn-on-Sea							12 37	X1238	12 22X	12 50										—	2 27
Talwrn Bach Halt							—	12 43	—	—										—	2 32
Llanbedr and Pensarn							—	X1247	1 0	X 1 20										—	2 35
Llandanwg Halt							—	12 50	—	—										—	2 37½
Harlech							12 56	1X 0	1 30X	1 50										2 42½	X243½
Tygwyn Halt							—	1 5	—	—										—	2 48½
Talsarnau							—	1 8	S	T										—	2 51½
Llandecwyn Halt							—	—	—	—										—	2 56
Penrhyndeudraeth							—	1 13½	2 15	X 2 40										— X	2 59½
Minffordd							—	1 17½	—	—										—	3 3½
Granite Quarry Sidings							—	—	—	—										—	—
PORTMADOC	Z	11 55					1 22	X1 27	2 55	—										3 8	—
Wern Siding	—						—	—													
Black Rock Halt	12 2	12 4					—	1 34½													
Criccieth	12 7½	X1212					1 38	1 39	Not to convey more than 25 wagons from Barmouth.												
Afon Wen	1217½	X1222			— X	1 20	1 44½	X1 48										—	2 55	Runs Saturdays, July 9th and 16th, and each week day from July 18th, to September 10th, inclusive.	
Penychain Halt	—	12 25			—	—	—	1 51										—	—		
Abererch	—	1228½			— ¶	—	—	1 54										—	3 0½		
Pwllheli (East) (Double Line)	—	—			—	—	—	—										—	—		
PWLLHELI (Double Line)	12 34	—			1 30	—	2 0	—										3 6	—		

¶—Calls to set down passengers on notice to guard.
‡—On Saturdays, Barmouth Junction 12.1, 12.2 p.m., Barmouth arr. 12.7 p.m.; not advertised.
§—Advertised 5 minutes later. N—To work 1.30 p.m. Barmouth to Chester.
U—On Saturdays, until September 3rd, inclusive, will leave Machynlleth empty at 12.20 p.m. for Dovey Junction. Passengers travel by 9.10 a.m. ex Birmingham.
Z—On Saturdays, August 6th, 13th, 20th, and 27th, will start at 11.56 p.m. Y—To work 2.35 p.m. Barmouth to Chester.

Down Trains. MACHYNLLETH, BARMOUTH, PORTMADOC and PWLLHELI. Week Days.	B Passenger. arr.	B Passenger. dep.	D Empty Train. (Runs on Saturday, Aug. 6th only.) arr.	D dep.	K Pilot. SO dep.	B Pass. SO (Will not run after August 27th.) dep.	A Passenger. 1.17 p.m. ex Ruabon SX (Will not run after Sept. 9th.) arr.	A dep.	K Pilot. SX dep.	A Passenger. 1.30 p.m. ex Ruabon. SO arr.	A dep.	B Passenger. 3.8 p.m. ex Dolgelley SX (Commences September 12th.) arr.	B dep.	B Passenger. arr.	B dep.	G L.E. SO Z dep.	B Passenger. 2.40 p.m. ex Ruabon. SO arr.	B dep.	B Passenger. arr.	B dep.		B Passenger. SO RR (Not advertised.) arr.	B dep.
STATIONS.	p.m.	p.m.	p.m.	p.m.	p.m.	p.m.	p.m.	p.m.	p.m.	p.m.	p.m.	p.m.	p.m.	p.m.	p.m.	p.m.	p.m.	p.m.	p.m.	p.m.		p.m.	p.m.
MACHYNLLETH	§	1† 5	—	1† 5																2†50		—	2†50
Dovey Junction	1†12	T1 37	1†12	2† 5															2†57	3 35		2†57	3 57
Gogarth Halt	—	V	—	—															—	—		—	—
Abertafol Halt	—	V	—	—															—	3 44½		—	—
Penhelig Halt	—	1 49	—	—															—	3 49		—	—
Aberdovey	1 52	1 53	—	—															3 52	3 53		4 11X	4 13
Towyn	—	2 0	2†25	—															3 59	X4 9		4 20	4 25
Tonfanau Quarry Siding	—	—																	—	—		—	—
Tonfanau	—	2 5																	—	4 15		—	4 30
Llangelynin Halt	—	—																	—	4 19		—	—
Llwyngwril	—	2 13																	4 22	4 25		4 38	4 39
Stop Board	—	—																	—	—		—	—
Fairbourne	—	2 21																	—	4 33		—	4 47
Dolgelley Branch Loop	—	—																	—	—		—	—
Barmouth Junction	2 23½	X2 29					3 15	3 16		3 29	X3 36	3 29	X3 36			4‖19	4 32	4 33	4 36	4 39		4 50	4 54
BARMOUTH	2 34	X2 37					3 21	X3 26		3 41	3 45	3 41	—	—	4 8	4‖25	4 38	X4 43	4 44	—		4X59	—
Llanaber Halt	—	V					—	—		—	—			—	4 12		—	—					
Talybont Halt	—	V					—	—		—	—			—	4 16½		—	—					
Dyffryn-on-Sea	—	2 46					Y3	34		Y3X	53			—	X4 20		Y4	51					
Talwrn Bach Halt	—	V					—	—		—	—			—	4 25		—	—					
Llanbedr and Pensarn	2 52	X2 53					Y3X	40		Y3	59			—	4 28		Y4	57					
Llandanwg Halt	—	V					—	—		—	—			—	4 30½		—	—					
Harlech	3 0	3 1	K Goods. SO				3 47	3 48		4 6	4 7	K Goods. SX		— X	4 36		5 4	X5 6					
Tygwyn Halt	—	V					—	—		—	—			—	4 41		—	—					
Talsarnau	—	3 7					Y3	54½		Y	—			—	4 44		Y5	12½					
Llandecwyn Halt	—	—	arr.	dep.			—	—		—	—	arr.	dep.	—	4 48½		—	—					
Penrhyndeudraeth	—	X3 13			3 26		Y4	X0	4 11	Y4	20			4 51½	X4 53½		Y5	18					
Minffordd	—	3 17	p.m.	p.m.	—		—	—	—	—	—	p.m.	p.m.	—	4 57½		—	—					
Granite Quarry Sidings	—	—			—		—	—	—	—	—				—		—	—					
PORTMADOC	3 22	3 26	X	4 2	3 40		4 8	4 9	4 25	4 28	X4 31	—	4 45	5 2	5 6		5 26	5 27					
Wern Siding	—	—	—	—			—	—		—	—	—	—	—	—		—	—					
Black Rock Halt	—	3 33½	—	—			—	—		—	—	—	—	—	5 13½		—	—					
Criccieth	3 37	3 38	4 15	X4●50			4 17½	X4 20		4 39½	X4 41	5 0	X●5 40	5X17	5 18½		5 35½	X5 37					
Afon Wen	3 43½	X3 48	5 0	—		4 20	—	—		4	46	5 50	—	5 24	X5 33		5C42½	5S45					
Penychain Halt	—	—				—	—	—		—	—			—	5 36		—	—					
Abererch	—	3 53½				—	—	—		—	—			—	5 39½		—	—					
Pwllheli (East) } Double	—	—				—	—	—		—	—			—	—		—	—					
PWLLHELI } Line.	3 59	—				4 30	4 35	—		4 55	—			5 45	—		5 55	—					

V—Will call each week day, July 4th to July 15th, and after September 10th.
§—Will leave Machynlleth at 12†20 p.m. on Saturdays to September 3rd.
Y—Calls to set down passengers from beyond Ruabon on notice to guard at Barmouth.
Z—SO during August only, to work 5.50 p.m. Barmouth to Ruabon.

No. 16 41

Down Trains. MACHYNLLETH, BARMOUTH, PORTMADOC and PWLLHELI. Week Days.	A Passenger 3.5 p.m. ex Ruabon. arr.	A dep.	B Pass. 3.12 p.m.ex R'bon dep.	B Pass. dep.	B Passenger. arr.	B dep.	B 5.25 p.m. Auto ex Dolgelley. (Will not run on Saturdays, August 6th, 13th, 20th, and 27th.) arr.	B dep.	G L.E. R dep.	K Goods or L.E. SX dep.	B Passenger. arr.	B dep.	B 6.38 p.m. Auto. ex Dolgelley. dep.	B Passenger. arr.	B dep.	K Pilot. SX dep.	K Goods. SX arr.	K dep.	A Passenger 5.14 p.m. ex Birkenhead. FSO arr.	A dep.	B Pass. (5.14 p.m. Birkenhead.) dep.	B Pass. SO dep.	
STATIONS.	p.m.	p.m.	p.m.	p.m.	p.m.	p.m.	p.m.	p.m.	p.m.	p.m.	p.m.	p.m.	p.m.	p.m.	p.m.	p.m.	p.m.	p.m.	p.m.	p.m.	p.m.	p.m.	
MACHYNLLETH					—	4‖15								—	5‖15								
Dovey Junction					4‖22	4 38								5‖22	6T 0								
Gogarth Halt					—	—								—	6 4½								
Abertafol Halt					—	U								—	6 10½								
Penhelig Halt					—	4 50								—	6 15								
Aberdovey					—	4 55								6 18	X6 22								
Towyn					—	5 2								—	X6 29								
Tonfanau Quarry Siding					—	—								—	—								
Tonfanau					—	5 7								—	6 34								
Llangelynin Halt					—	—								—	6 39								
Llwyngwril					—	5 15								—	6 43								
Stop Board					—	—								—	—								
Fairbourne					—	5 23								—	6 51								
Dolgelley Branch Loop					—	—								—	—								
Barmouth Junction	4 46	4 47	5 17		5 26	X5 27	5 45	X5¶50	6‖34				7 0	6 53½	6 54			6X45	8 19	X8 20	N 9 0		
BARMOUTH	4 53	5 0	5 22		5 32	X5 40	5¶55	—	6‖40			6X42	7X 5	6 59	X7V15		6 52	X 7 30	8 25	8 30	9 5		
Llanaber Halt	—	—			—	5 44						6 46		—	7 19		—	—	—	—			
Talybont Halt	—	—			—	5 49					—	6 50½		—	7 23½		—	—	—	—			
Dyffryn-on-Sea	Y5	8			—	X5 53					—	6 54		7 26½	7 27½		7 47	X 8 5	Y	—			
Talwrn Bach Halt	—	—			—	5 58					—	6 59		—	7 32½		—	—	—	—			
Llanbedr and Pensarn	Y5	X14			—	6 1½					V	X7 2		7 35	7 36		8 15	●8 50	Y	X			
Llandanwg Halt	—	—			—	6 4					—	7 4½		—	7 38½		—	—	—	—			
Harlech	5 21	X5 22			6 9	X6 12½					—	7 10½		7 43½	X7 46		9 0	X 9 25	8 50	8 51			
Tygwyn Halt	—	—			—	6 17½					—	7 15½		—	7 51		—	—	—	—			
Talsarnau	Y5	29			—	6 20½					—	7 18½		—	7 54		—	—	Y	—			
Llandecwyn Halt	—	—			—	6 25					—	7 23		—	—		—	—	—	—			
Penrhyndeudraeth	Y5	35			—	6 28½					7 26X	7 29		7 59	8 0		L	—	Y 9	X4			
Minffordd	—	—			—	6 32½					—	7 33		—	8 4		—	—	—	—			
Granite Quarry Sidings	—	—			—	—					—	—		—	—	8 37	—	—	—	—			
PORTMADOC	5 43	X5 45			6 37	X6 39				7X15	7 37½	7 39		8 8½	X8 10	8 45	10 0	—	9 12	9 14			
Wern Siding	—	—			—	—				—	—	—		—	—				—	—			
Black Rock Halt	—	—			—	—				—	—	7 46½		—	—				—	—			
Criccieth	5 53½	5 55			—	6 48				§7X28	7 50X	7 51		—	8 19				9 22½	9 23			
Afon Wen	—	—		6 20	6 53½	X6 56				7 40	7	56		8 24	X8 28				9X	28		9 50	
Penychain Halt	—	—		—	—	6 59					—	—		—	8 31				—	—		—	
Abererch	—	—		—	—	7 2½					—	—		—	8 34½				—	—		9 55	
Pwllheli (East) } Double	—	—		—	—	—					—	—		—	—				9X	36		—	
PWLLHELI } Line.	6 10	—		6 30	7 8	—					8 5	—		8 40	—				9 38	—		10 1	

N—Starts from Ruabon on Fridays and Saturdays.
R—To work 7.15 p.m. Barmouth to Chester.
Y—Calls to set down passengers from beyond Ruabon on notice to guard at Barmouth.
¶—Advertised 4 minutes earlier.
U—Calls to set down passengers from Welshpool or beyond on notice to guard at Dovey Junction.
V—Crosses 6.10 p.m. Pwllheli on Wednesdays, July 20th to September 7th, inclusive.
§—Arrives Criccieth 7.28 p.m.

42 No. 16

MACHYNLLETH, BARMOUTH, PORTMADOC and PWLLHELI.	Down Trains. Week Days. B Passenger. arr.	B dep.	B Passenger. SO arr.	B dep.		Down Trains. Sundays. B Passenger. arr.	B dep.	B L.M.S. Pass. (July 24th to September 11th, inclusive.) dep.	B Passenger. (July 24th to September 11th, inclusive.) arr.	B dep.	K Goods. (Runs on August 7th, 14th, 21st, and 28th only.) arr.	K dep.	K Goods. RR arr.	K dep.	A 9.48 a.m. ex Birkenhead. Passenger. arr.	A dep.	B Pass. dep.	B L.M.S. Pass. (July 24th to September 11th inclusive.) dep.	B Passenger ex Birmingham. arr.	B dep.	B Passenger. (July 24th to September 11th, inclusive.) arr.	B dep.	
STATIONS.	p.m.	p.m.	p.m.	p.m.		a.m.	a.m.	a.m.	p.m.	p.m.	a.m.	a.m.	p.m.	p.m.	p.m.	p.m.	p.m.	p.m.	p.m.	p.m.	p.m.	p.m.	
MACHYNLLETH	§	7‖40	—	10 10		—	8 25					9 0							—	2 30			
Dovey Junction	7‖47	8 20	10	16		8	31				—	—							2	36			
Gogarth Halt	—	V	—	—		—	—				—	—							—	—			
Abertafol Halt	—	V	—	—		—	—				—	—							—	—			
Penhelig Halt	—	8 33	—	—		—	8 43				—	—							—	2 48			
Aberdovey	—	8 37	10 31	X1036		—	8 47				9 30	9 43							—	2 52			
Towyn	—	8 44	10 42	—		8 53	8 54				9C55	10T0							—	2 59			
Tonfanau Quarry Siding	—	—				—	—				—	—							—	—			
Tonfanau	—	8 49				—	8 59				—	—							—	—			
Llangelynin Halt	—	8 53				—	—				—	—							—	—			
Llwyngwril	C	8X57				9 6	9 7				10 51	11 5							3 10	X3 12			
Stop Board	—	—				—	—				P	—							—	—			
Fairbourne	—	9 5				9 14	9 15				C	R							—	3 20			
Dolgelley Branch Loop	—	—				—	—				—	—							—	—			
Barmouth Junction	9 8	9 9				—	9 18				11 25	11 35			12 47	12 48			—	3 23			
BARMOUTH	9 14	9 25				9 23	—				11 42	12 0			12 53	—	1 30		3 28	3 35			
Llanaber Halt	—	9 29									—	—					1 34		—	3 39			
Talybont Halt	—	9 33½									—	—					1 39		—	3 44			
Dyffryn-on-Sea	9 36½	X9 39									12 17	12 30					1 43		—	3 48			
Talwrn Bach Halt	—	9 44									—	—					1 48		—	3 53			
Llanbedr and Pensarn	—	9 47									12 40	12 50					1 51		—	3 56			
Llandanwg Halt	—	9 49½									—	—					1 54		—	3 59			
Harlech	9 54½	9 55									1 0	1 15					2 0		4 5	—			
Tygwyn Halt	—	10 0									—	—											
Talsarnau	—	10 3									—	—											
Llandecwyn Halt	—	10 7½									—	—											
Penrhyndeudraeth	1010½	1011½									1 35	1 46											
Minffordd	—	1015½									—	—											
Granite Quarry Sidings	—	—									—	—											
PORTMADOC	10 20	10 21							—	12 10	2 0	—		2 45							—	5 0	
Wern Siding	—	—							—	—			—	—							—	—	
Black Rock Halt	—	—							—	—			—								—	5 8	
Criccieth	—	10 30							—	12 19			2 58	3 12							—	5 12½	
Afon Wen	—	—						11 39	1224½	12X28			3 20	—				3 29			5 18	X5 20	
Penychain Halt	—	—						—	—	12 31								—			—	5 23	
Abererch	—	—						1144½	—	1234½								3 34½			—	5 26½	
Pwllheli (East) { Double	—	—						—	—	—								—			—	—	
PWLLHELI { Line.	10 45	—						11 50	12 40	—								3 40			5 32	—	

RR—On August 7th, 14th, 21st, and 28th only.
C—Will cross 8.40 p.m. ex Barmouth on Saturdays, July 30th to September 3rd.
V—Calls to set down on notice to the guard.
§—Leaves Machynlleth at 7X21 p.m. on Saturdays.

No. 16 43

Train Services (continued)

from Portmadoc, ex-Cambrian 0-6-0 tender engines worked the train.

A light engine would run to Chwilog, setting back in the cattle dock on the Afonwen side of the level crossing to await a train of empty coaching stock. The waiting light-engine would then return to the main line to join the stock enabling the release of the train engine which then remained in the platform until after the excursion had departed.

When the passenger had cleared Afonwen and released the single line token, the light-engine entered the section and followed the excursion to Portmadoc, where it went on shed. The return working that evening followed the same procedure.

Coach Workings

Apart from the Dolgellau to Barmouth shuttle, which was worked by steam railcars after amalgamation with the GWR in 1922 and which survived the duty until the early 1930's only to be replaced by standard GWR design Autocoaches, most of the passenger working on the coast line since the amalgamation has been with GWR corridor coaches.

The usual formations have been three or four coach sets, strengthened as required for certain journeys. Details of pre-war workings are not available, but the circuit diagrams for local sets for 1952 and 1954 are to hand. The Central Wales district working book indicated the make up and distribution of all stock involved in the normal schedule services. Along the coast, circuits 24 to 39 comprised of 5 four coach sets of which one set was spare at Barmouth and three were on two-day workings; 3 five coach sets, one set involved in a five day working circuit and one on a two day working; 2 three coach sets, 5 two coach sets and 2 one coach sets. There were five vehicles spare during the week, 2 thirds at Barmouth and Pwllheli and a brake composite at Portmadoc.

Locomotive Workings

Following the 1948 nationalization, the London Midland Region assumed responsibility for running some trains to Pwllheli rather than changing engines, causing some loss of work to both Portmadoc and Pwllheli sheds, although the former lost more work. Bangor 2-6-4T's ran round trains at Afonwen and worked to Pwllheli, rather than waiting in the Up goods loop or the turntable road for the stock to return. In the summer season, from the other direction, the LM Region worked a class 3 Stanier design 2-6-2T light engine to Portmadoc where the two or three coaches which constituted 'The Welshman' stock was drawn out of the sidings and into the Down platform line. According to J. M. Dunn, who travelled on the first engine to work the duty and who took photographs of the train at Portmadoc, the locomotive was number 73, still in LMS black livery, and three coaches. The journey was uneventful, and at Afonwen, the stock from Pwllheli was attached making a six coach formation. It seems ironic that such a heavy formation was considered acceptable, for Mr. Dunn had been campaigning for many years to

Locomotive workings (continued)

have the 2-6-2T's replaced by 2-6-4T's because the smaller engines could not cope with the more usual four coach trains that were used for the services. Be that as it may, the normal formation was a two coach set for Euston from Portmadoc and Pwllheli.

The older locomotives gradually disappeared from the line, their place being taken by slightly less venerable replacements. The most common types were the outside framed 4-4-0 tender engines, formerly 32XX and latterly 90XX universally known as 'Dukedogs', classes ideal for the working of the line. Most of the members of the class worked over the coast line at some period in their lives, and some, like 9000, 9004, 9011, 9014 and 9017 rarely strayed from the coast apart from visits to Oswestry, Stafford Road or Swindon for overhaul. The former Cambrian 0-6-0's with their Swindon boilers remained visible until the mid 1950's, and 849, 892 and 894 spent most of their lives at Pwllheli or Portmadoc on passenger and freight workings. The 45XX tanks were in evidence although somewhat limited because of their water capacity and were replaced by their cousins, the 4575 class. These also worked passenger and freight duties, but contained the latter to short distance workings nominally between Portmadoc and Pwllheli.

Penmaenpool shed, which strictly speaking is outside the realms of this work, is nevertheless drawn into the picture by virtue of the fact that the Dolgellau to Barmouth shuttle service featured regularly on the coast line. Pre and post war working was in the hands of 0-4-2T's working an auto train, which had superceded a steam railcar in the early 1930's. In 1957 it was still possible to find a locomotive with original GWR markings on the tank sides, some nine years after nationalization.

The Dean 0-6-0's had worked spasmodically on the coast for many years but were less well known than their modern Collett equivalents, engines which were popular with traincrews. These locomotives were normally attached to Machynlleth shed but frequently found themselves boarded overnight at either of the two depots at the northern end of the line.

Through workings from Chester were usually in the hands of 43XX class 2-6-0 tender engines and although they rarely lodged at Pwllheli, could be seen during the day whilst waiting their return working.

The Manor class 4-6-0's were the largest type permitted to work over the line despite their low axle load and were prohibited from entering certain sidings. Nevertheless they became frequent visitors from about 1951 onwards, and could sometimes be seen on the Up 'Cambrian Coast Express' from Pwllheli, usually on Saturdays. From 1953 onwards, they could also be found working the Saturdays Only working from Butlin's Camp to Swansea via Dovey Junction and Aberystwyth. Pwllheli shed's proudest moment was probably on Saturday 10th August 1963 when at 4.35p.m. it was host to the four Manor's which were involved with royal train workings. For the record, engines were

44

Up Trains. PWLLHELI, PORTMADOC, BARMOUTH and MACHYNLLETH. Week Days.

No. 16

Miles from Pwllheli M.	C.	STATIONS.	Ruling Gradient 1 in	Time Allowances for Freight Trains—see page 2. Point to Point Allowances: Allow for Stop.	Allow for Start.	"C" Head Lamp Trains	"D" Head Lamp Trains	Express Trains	Ordinary Trains	G L.E. dep.	K Pilot. dep.	B Passenger and Mail. arr.	B Passenger and Mail. dep.	B Pass to R'bon dep.	K Goods. arr.	K Goods. dep.	B Pass. dep.	B Passenger. arr.	B Passenger. dep.	K Goods. WSO arr.	K Goods. WSO dep.	B Passenger to Paddington. SO arr.	B Passenger to Paddington. SO dep.	B Auto. to Dolgelley. dep.
				Mins.	Mins.	Mins.	Mins.	Mins.	Mins.	a.m.	a.m.	a.m.	a.m.	a.m.	a.m.	a.m.	a.m.	a.m.	a.m.	a.m.	a.m.	a.m.	a.m.	a.m.
..	..	bPWLLHELI { Double Line.	100 F.							5‖45		—	6 5				6 45							
0	50	bcPwllheli (East)	631 F.		1					—		—	—				—							
1	68	Abererch								—		—	—				—							
3	32	Penychain Halt								—		—	—				—							
4	48	cAfon Wen	100 F.	1	1			8	13	5‖55		6 15	X6 20				6X55							
7	71	cCriccieth	64 R.	1	1				11			6 25½	6 27											
9	27	Black Rock Halt										—	—											
11	9	Wern Siding										—	—											
11	19¾	Stop Board										—												
12	63	cPORTMADOC	54 R. 50 F.	1	1			22	13			6 35	X6 40		—	4 30								
14	37	Granite Quarry Siding										—	—		—	—								
15	7	dMinffordd	50 R.	1	2				6			—	6 45		—									
15	49¾	Stop Board										—	—		—	P								
16	18	cPenrhyndeudraeth	50R. 65F	1	1			13	5			—	6 50		4 49	5 0		Will not run on Saturdays nor when Towyn Schools are closed.		Except Wednesdays, when school train runs. On Saturdays, must not convey Departmental traffic, but must be limited to S.T.'s and through loads for Towyn, Aberdovey, and Machynlleth.		☞ Will run on Saturdays July 16th to Sept. 10th, inslusive.		
17	2	Llandecwyn Halt										—	—		—	—	G							
18	22	Talsarnau	102 F.	1	1				10			—	6 56		5 10	5 20								
19	10	Tygwyn Halt										—	—		—	—	L.E.							
21	52	cHarlech	660 R.	1	1				8			—	7 2		5 30	5 45								
23	48	Llandanwg Halt										—	N		—	—	R							
24	20	cLlanbedr and Pensarn	75 F.	1	1				8			—	7 9		5 55	6 5								
24	77	Talwrn Bach Halt										—	—		—	—	dep.							
27	15	cDyffryn-on-Sea	75 F.	1	1				10			—	7 15		6 15	6 25	a.m.							
28	26	Talybont Siding										—	—		C	R								
28	28	Talybont Halt										—	7 19		—	—								
30	51	Llanaber Halt										—	—		—	—								
32	15	cBARMOUTH	75 F.	1	1			43	13			7 25	X7 28	7 40	6 42	X•7 53	8‖10	—	—			— X	9¶17	X9¶27
33	65	cBarmouth Junction	222 F.	1	1			5	5			7 33	7 34	7 45	8 0	—	8‖16	—	8 20	—	X8 45	9 22	X9 23	9 32
..	..	Dolgelley Branch Loop										—	—					—	—	—	—	—	—	
34	73	Fairbourne	76 R.	1	2				3			—	7 37					—	8 23	—	—	—	9 27	
37	46	cLlwyngwril	55 R.	1	1				9			— X	7 45½					8C31X	8T35	9 0	X9 11	—	9 36	
39	41	Llangelynin Halt	—									—	7 50					—	8 39½	—	—	—	—	
41	56	dTonfanau	75R.F.	1	1				14			—	7 55					—	8 44½	—	—	—	9 44	
41	67	Tonfanau Quarry Siding									18	—	—					—	—	—	—	—	—	
44	8	cTowyn	120 F.	1	1			30	5		25	7 59½	X8 1					8X50	—	9X34	•1030	9 48½	X9 51	
47	63	cAberdovey	70 R.	1	2				12			—	X8 8							10•44	X1143	9 57	9 59	
48	57	Penhelig Halt										—	8 11							—	—	—	10 3	
50	36	Abertafol Halt										—	8 16							—	—	—	—	
52	11	Gogarth Halt										§	—							—	—	—	—	
53	62	cDovey Junction	98 F.	1	1				15			8X25	8‖40							12	0	10 16	10‖23	
57	54	cMACHYNLLETH	132 R.	1				33	9			8‖47	—							12X10	—	10‖30	—	

N—Calls on Mondays only, not advertised. R—Engine of 7.0 a.m. ex Dolgelley.
¶—Advertised 2 minutes earlier. §—Calls to set down on notice being given at Aberdovey.

Up Trains. PWLLHELI, PORTMADOC, BARMOUTH and MACHYNLLETH. Week Days.

No. 16

STATIONS.	B Passenger. arr.	B Passenger. dep.	B Passenger. SX arr.	B Passenger. SX dep.	A Passenger to Birkenhead. SO arr.	A Passenger to Birkenhead. SO dep.	G L.E. SO dep.	K Goods WSX arr.	K Goods WSX dep.	B Pass. SX dep.	B Pass. to Ruabon. SX dep.	B Passenger to Birmingham SO arr.	B Passenger to Birmingham SO dep.	A Passenger to Wrexham SX Ruabon SO arr.	A Passenger to Wrexham SX Ruabon SO dep.	B Passenger. arr.	B Passenger. dep.	B Passenger. arr.	B Passenger. dep.	B Passenger. arr.	B Passenger. dep.	B Passenger. SO arr.	B Passenger. SO dep.
	a.m.	a.m.	a.m.	a.m.	a.m.	a.m.	a.m.	a.m.	a.m.	a.m.	a.m.	a.m.	a.m.	a.m.	a.m.	a.m.	a.m.	a.m.	a.m.	a.m.	a.m.	a.m.	a.m.
PWLLHELI } Double Line.	—	7 55								9 5		—	9 5	—	9 35	—	10 15					☞ Runs July 16th to	Sept. 3rd inclusive.
Pwllheli (East)	—	—								—		—	—	—	—	—	—						
Abererch	—	8 0½								9 10½		—	9 10½	—	—	—	1020½						
Penychain Halt	—	8 4								—		—	—	—	—	—	10 24						
Afon Wen	8 6½	8 9								9 16		9 16X	9 19	—	—	1027X	10 32						
Criccieth	8 14½	8 15½										9 24½	9 26	9 50	9 51	1037½	X1039½						
Black Rock Halt	—	8 19										—	—	—	—	—	10 43						
Wern Siding	—	—										—	—	—	—	—	—						
Stop Board	—	—															—						
PORTMADOC	8 26	X8 28	—	8 55								9 34	9 35	9 59	10 0	10 50	—					—	10 53
Granite Quarry Siding	—	—	—	—								—	—	—	—							—	
Minffordd	—	8 33	—	9 0								—	—	—	—							—	10 58
Stop Board	—	—	—	—								—	—	—	—							—	—
Penrhyndeudraeth	—	8 37	—	9 5				Will also run on Wednesdays when the school train runs.				¶9	43½	10¶8X	10Y11							— X	11 2
Llandecwyn Halt	—	8 40½	—	9 8½								—	—	—	—							—	—
Talsarnau	—	8 45	—	9 13								¶9	49	¶10	16½Y							—	11 8
Tygwyn Halt	—	8 48	—	9 16								—	—	—	—							—	11 11
Harlech	—	8 53	—	9 22								9 55	X9Y58	10 22	10 23			Runs on Saturdays, July 9th and 16th, and daily from July 18th to September 10th, inclusive.		—	11 0	1115½	11 17
Llandanwg Halt	—	8 58	—	9 27½								—	—	—	—					—	11 5½	—	—
Llanbedr and Pensarn	—	9 1	—	9 31								¶10	5	¶10X	30					—	11 8½	1123½	11 25
Talwrn Bach Halt	—	9 3½	—	9 33½								—	—	—	—					—	11 11	—	—
Dyffryn-on-Sea	—	9 9	9 38X	9 42								¶10	X11	¶10X	36					1115½	X11C17	1130½	11 32
Talybont Siding	—	—	—	—								—	—	—	—					—	—	—	—
Talybont Halt	—	9 13	—	9 45								—	—	—	—					—	11 21	—	—
Llanaber Halt	—	9 17½	—	9 49½								—	—	—	—					—	11 26	—	—
BARMOUTH	9 21	X9 40	9 53	—	—	9Y58					10 15	1018½	10 25	1043½	10 50				10 40	11 30	—	11 40	11 43
Barmouth Junction	9 45	9 47	Runs July 18th to September 9th only.		10X 4	10 6		—	•10 5		10 20	10 31	10 32	10 56	10 57			1045X	10C47			11X48	11R55
Dolgelley Branch Loop	—	—						—	—													—	—
Fairbourne	9 50	9 53						—	—									10 50	10 52			— R	1158½
Llwyngwril	—	10 1½						10 20	10 35									—	11 0½			—	12 X7
Llangelynin Halt	—	10 6						—	—									—	11 5			—	—
Tonfanau	—	10 11						—	—									—	11 10			—	12 15
Tonfanau Quarry Siding	—	—						—	—									—				—	
Towyn	1015½	X1017						10•58	X11 35									1114½X	11‡20			1219½	12 21
Aberdovey	1023X	1023½						11 49	11 58									— X	1126½			12 27	12 28
Penhelig Halt	—	10 27						—	—									—	1129½			—	12 31
Abertafol Halt	—	10 32						—	—									—	—			—	—
Gogarth Halt	—	10 38						—	—									—				—	—
Dovey Junction	10 42	10V54					11‖13	12	15									1140½	11U43			12 42	X1‖45
MACHYNLLETH	11V 0	—					11‖20	12 25	—									11U49	—			1‖52	—

C—Will cross 10.8 a.m. ex Dovey Junction on Wednesdays, July 20th to September 7th, inclusive. R—Advertised 6 minutes earlier.
U—Saturdays excepted to September 3rd, passengers for Machynlleth to go forward by 11.5 a.m. ex Aberystwyth. Engine runs light on Saturdays to Sept.3rd, inclusive, Dovey Jct. dep. 11‖51, Machynlleth arr. 11.58 a.m.
V—SX Dovey Junction to Machynlleth.
Y—Advertised 3 minutes earlier. ‡—On Saturdays will leave Towyn at 11.16 and arrive Aberdovey at 11X22 a.m. ¶—Calls to pick up passengers for Ruabon or beyond.

45

46

Up Trains. PWLLHELI, PORTMADOC, BARMOUTH and MACHYNLLETH. Week Days.

No. 16

STATIONS.	B Passenger. arr.	B Passenger. dep.	K Pilot. dep.	A Pass. Birkenhead SO dep.	B Passenger. arr.	B Passenger. dep.	B Auto to Dolgelley. arr.	B Auto to Dolgelley. dep.	G L.E. SO dep.	B Passenger. SO arr.	B Passenger. SO dep.	K Pilot. arr.	K Pilot. dep.	B Passenger to Birmingham. SO arr.	B Passenger to Birmingham. SO dep.	K Goods. arr.	K Goods. dep.	B Pass. SO dep.	B Pass. SX dep.	B Passenger. arr.	B Passenger. dep.	B Passenger to Birkenhead. arr.	B Passenger to Birkenhead. dep.
	a.m.	a.m.	a.m.	p.m.	p.m.	p.m.	p.m.	p.m.	p.m.	a.m.	a.m.	p.m.	p.m.	a.m.	a.m.	a.m.	a.m.	a.m.	p.m.	p.m.	p.m.	p.m.	p.m.
PWLLHELI } Double Line.	—	10 40									9†50			—	11 50			11 35	12 10			—	12 10
Pwllheli (East)	—	—												—	—	—	11 20	—	—			—	—
Abererch	—	—												—	—	—	—	1140½	1215½			—	1215½
Penychain Halt	—	—												—	—	—	—	—	—			—	—
Afon Wen	10 50	X1052								10	0			11	59	11 33X	•1230	11 46	12X21			1221X	12 23
Criccieth	—	10 58								10† 5	11 45			12 5	X12 9	12 43X	12 52					—	12 29
Black Rock Halt	—	—								—	1148½			—	—	—	—					—	—
Wern Siding	—	—	N							—	—			—	—	—	—					—	—
Stop Board	—	—								—	—			—	—	P	—					—	—
PORTMADOC	11 6	X11 7	11 20	☞ Will not run after Sept. 3rd.						1155½	X1157	—	12 10	12 17	12 19	1 10	—		Commences September 12th.			12 37	X1238
Granite Quarry Siding	—	—	11 27							—	—			—	—							—	—
Minffordd	—	11 12								—	12 2	12 18	X•1 50	—	12 24							—	12 43
Stop Board	—	—								—	—	P		—	—							—	—
Penrhyndeudraeth	—	11 16								—	12 6	2 0	—	—	12 28							—	12 47
Llandecwyn Halt	—	1119½								—	—			—	—							—	—
Talsarnau	—	11 24								—	1211½			—	12 33							—	1252½
Tygwyn Halt	—	11 27								—	1214½			—	—							—	—
Harlech	1131½	X1133								12 19	12 20			1238½	12 40							12 58	X1 0
Llandanwg Halt	—	Z								—	1225½			—	—							—	—
Llanbedr and Pensarn	—	11 42								1228½	12 29			1246½	X1248							1 6½X	1 7½
Talwrn Bach Halt	—	Z								—	12 32			—	—							—	—
Dyffryn-on-Sea	—	11 49							V	1236½	X1240			12 54	X1257							1 13	1 14
Talybont Siding	—	—								—	—			—	—							—	—
Talybont Halt	—	—								—	12 45			—	—							—	—
Llanaber Halt	—	—								—	12 50			—	—							—	—
BARMOUTH	11 58	—		12X10	—	12 20	—	12 30	12‖36	12 54	—			1 5	1 15							1 22	X1 30
Barmouth Junction				12 15	12 25	12 28	12 35	12 36	12‖42					1 21	1 22							1 36	1 37
Dolgelley Branch Loop					—	—				Runs on Saturdays, August 6th, 13th, 20th, and 27th only.				Runs August 6th, to September 10th inclusive								Runs daily to September 10th, also on Saturdays, September 17th and 24th.	
Fairbourne					—	12 32														Runs on Saturday, August 6th only.			
Llwyngwril					—	X1241																	
Llangelynin Halt					—	12 45																	
Tonfanau					—	12 50																	
Tonfanau Quarry Siding					—	—																	
Towyn					1254½	12 57															2 35		
Aberdovey					1 3	X1 8														2T 41	2 43		
Penhelig Halt					—	1 11														—	2 47		
Abertafol Halt					—	1 16														—	—		
Gogarth Halt					—	1 22														—	—		
Dovey Junction					1 26	1‖45														3 0	—		
MACHYNLLETH					1‖52	—																	

N—On the Wednesdays and Saturdays when this schedule is impracticable, Station Master to advise Quarry Manager and arrange a special pilot trip later in the day.
V—Engine of 8.57 a.m. ex Chester.
Z—Calls to set down passengers on notice to guard.

Locomotive Workings (continued)

7822 *Foxcote Manor*, now preserved at Llangollen; 7819 *Hinton Manor*, now on the Severn Valley Railway and visitor to the Cambrian Coast in the summer 'Return to Steam' workings with the 'Cardigan Bay Express'. The third Manor was 7827 *'Lydham Manor*, preserved and working on the Torbay Steam Railway, and finally 7828 *Odney Manor*, restored to steam in 1988 on the Gwili Railway.

With the removal of the older classes from the stock-books, from about 1953 onwards, British Railways standard classes gradually appeared on the coast line. Portmadoc shed was allocated 2-6-0 tender engines of the 78XXX class, numbers 78002 and 78003 being allocated to the shed new where they took over workings from the ex-Cambrian 0-6-0's, now consigned to valhalla. They were received with mixed feelings, the enclosed tender cap providing much needed protection. Tender first working was not unknown, despite the practice of turning locomotives at Pwllheli for return workings up the coast. It was rare indeed to find a Western Region tank engine working bunker first for journeys of more than a few miles on this line.

When the 4-4-0's were being withdrawn, their places were taken by Standard class 3 2-6-2T's including 82009/32/3/4 which worked a three day diagram, spending one night at each of the coast sheds. Working both passenger and freight turns, they seemed to suffer the same problems the 45XX class had experienced ten years earlier so their workings were confined to shorter haul trips and subsequently replaced by the larger 80XXX class 2-6-4T. Locomotives 80096/8/104 appeared on the coast, with 82032 transferring to Bangor, a move regarded as retrograde by the 6H men. The 78XXX class engines were replaced by the LMS designed 465XX types and were preferred by most crews. It was left to a 465XX from Penmaenpool to confine to memory the Dolgellau-Barmouth shuttle in 1965.

British Railways standard class 4-6-0 locomotives were regular performers on the line towards the end of steam but were nearly always based at Machynlleth shed. They were popular with train crews although the reverser, which was inclined to be stiff, was considered to be a 'rupture inducer' to drivers.

With the decline of steam, passenger workings were transferred to DMU's which came with a dramatic suddenness, taking over services from January 18th 1965. Freight traffic was on the decline and gradually disappeared, surviving business being entrusted to class 24 Bo-Bo locomotives until final demise.

The assortment of DMU's working the line provided the only variety, the workings of which were part of intensive rostering based at Chester, Crewe and Shrewsbury, although in 1985, units might be overnight at Pwllheli on Monday, Chester on Tuesday and Shrewsbury on Wednesday, repeating the cycle the following night. (Chester circuit 121/2/3). The older DMU's were replaced overnight by new 'Sprinter' units in 1986 although casualties ensured that the older designs did not disappear altogether.

Up Trains. PWLLHELI, PORTMADOC, BARMOUTH and MACHYNLLETH. Week Days.

STATIONS.	A Passenger to Birmingham. SO arr.	A dep.	G L.E. SO dep.	B Passenger to Birkenhead. arr.	B dep.	B Passenger. arr.	B dep.	K Goods. SO arr.	K dep.	B Passenger. arr.	B dep.	G L.E. FSO dep.	F Goods. arr.	F dep.	K Goods SX arr.	K dep.	G L.E. SO dep.	B Pass. SO dep.	B Passenger. arr.	B dep.	B Passenger and Mail. arr.	B dep.
	p.m.	p.m.	p.m.	p.m.	p.m.	p.m.	p.m.	p.m.	p.m.	p.m.	p.m.	p.m.	p.m.	p.m.	p.m.	p.m.	p.m.	p.m.	p.m.	p.m.	p.m.	p.m.
PWLLHELI (Double Line)				—	12 55					—	1 35				—	—			—	3 30	—	4 0
Pwllheli (East)				—	—			—	1 7						—	2 0	2‖15		—	—	—	—
Abererch				—	1 0½						1 40½				—		—		—	3 35½	—	4 5½
Penychain Halt				—	—			—	—	—	1 44				—	—	—		—	3 38½	—	4 8½
Afon Wen				D	—			1 19	X	1 47	X1 50				2 12	2 25	2‖25	3 0	3 42	X3 46	4 11½	4 13
Criccieth				1 11	1X13					—	1 56½				2 38	2 55		3 6	—	3 52	4 18½	4X19½
Black Rock Halt				—	—					—	2 0				—	—		—	—	—	—	4 23
Wern Siding				—	—					—	—				—	—		—	—	—	—	—
Stop Board				—	—					—	—				P	—		—	—	—	—	—
PORTMADOC				1 21	X1 23					2 7	2 10		—	2 20	3 15	—		3 15	4 0	—	4 30	X4 37
Granite Quarry Siding				—	—					—	—		—	—							—	—
Minffordd				—	1 28					—	2 15		—	—							—	4 42
Stop Board				—	—					—	—		P	—							—	—
Penrhyndeudraeth				—	1 32					—	X2 19		2 39	X 3 15							4 46	X4 52
Llandecwyn Halt				—	—					—	2 23		—	—							—	—
Talsarnau				—	1 37½					—	2 28		C	O							—	4 58
Tygwyn Halt				—	1 40½					—	2 31		—	—							—	5 1
Harlech				1 45X	1 46					2 36	X2 44		C	O							5 5½	X5 6
Llandanwg Halt				—	1 51½					—	2 49½		—	—							—	5 11½
Llanbedr and Pensarn				—	1 54					2 51½	X2 54	N	3 39	X3V45							5 13½	X5 15
Talwrn Bach Halt				—	1 56½					—	2 56½		—	—							—	5 17½
Dyffryn-on-Sea				—	2 2			D Empty Auto.		3 1	3 2		3V55	X 4 22							5 22	5‡23
Talybont Siding				—	—			arr.	dep.	—	—		—	—							—	—
Talybont Halt				—	2 6			p.m.	p.m.	—	3 6		—	—							—	5‡27
Llanaber Halt			R	—	2 10½					—	3 10½		—	—							—	5‡31½
BARMOUTH	— X	2Z13	2‖13	2 14	X2 35	—	2 23	—	3† 5	3 14	X3 30	4‖ 0	4 37	X 5 0							5 35	X5 43
Barmouth Junction	2 18	2 20	2‖18	2 40	2 41	2 28	X2 30	3†10	—	3 35	X3 37	4‖ 6	5 7	X 5 27							5 48	X5 49
Dolgelley Branch Loop						—	—			—	—		—	—							—	—
Fairbourne						—	2 33			—	3 40½		—	—							—	5 53
Llwyngwril						—	2 42			—	3 49		—	—							—	6 2
Llangelynin Halt						—	2 46			—	3 54		—	—							—	—
Tonfanau						—	2 51			—	3 59		—	—							—	6 10
Tonfanau Quarry Siding						—	—			—	—		—	—							—	—
Towyn						2 55½	2 57			4 4	X4 5		5 59	●X6 35							—	X6 15
Aberdovey						3 3	3 4			4 11	X4 12		—	—							6 21X	6 23
Penhelig Halt						—	3 7			—	4 15		—	—							—	6 26
Abertafol Halt						—	3 12			—	4 20		—	—							—	Y
Gogarth Halt						—	3 18			—	4 26		—	—							—	Y
Dovey Junction						3 22	3‖50			4 30	X4 39		7	0							6 40	7†13
MACHYNLLETH						3‖57	—			4 45	—		7 10	—							7†20	—

A (Passenger to Birmingham): Runs August 6th, 13th, 20th, 27th only.
G (L.E. SO): Will not run on August 6th, 13th, 20th, or 27th.
K (Goods SO): On Saturdays, Sept. 17th and 14th, Engine, etc., will work a Special Goods from Afon Wen to Portmadoc.
G (L.E. SO) and B (Pass. SO): Will not run after Sept. 10th.

D—Calls commencing September 12th to pick up passengers.
CO—Calls for Live Stock only. N—Engine of 1.17 p.m. (FO to September 9th) and 1.30 p.m. SO ex Ruabon. R—Engine of 11.0 a.m. ex Ruabon.
V—On Saturdays this train must be limited to 20 wagons of loaded traffic, and will call at Llanbedr and Pensarn for Livestock only, and arrive Dyffryn at 3X50 p.m.
Y—Calls to set down on notice to guard at Aberdovey. Calls regularly on Fridays. Z—Advertised 3 minutes earlier. ‡—Advertised 2 minutes earlier.

No. 16 — 47

Up Trains. PWLLHELI, PORTMADOC, BARMOUTH and MACHYNLLETH. Week Days.

STATIONS.	A Passenger to Wrexham. SO arr.	A dep.	B Auto to Dolgelley. arr.	B dep.	G L.E. dep.	K Goods. SX arr.	K dep.	D E.C. SO dep.	B Passenger. arr.	B dep.	D E.C. SX dep.	B Pass. and Mail. dep.	A Passenger to Aberystwyth. arr.	A dep.	B Pass. dep.	G L.E. R dep.	B Passenger and Mail. arr.	B dep.	K Goods. arr.	K dep.	K Pilot. SX dep.	D E.C. SO dep.
	p.m.	p.m.	p.m.	p.m.	p.m.	p.m.	p.m.	p.m.	p.m.	p.m.	p.m.	p.m.	p.m.	p.m.	p.m.	p.m.	p.m.	p.m.	p.m.	p.m.	p.m.	p.m.
PWLLHELI (Double Line)								5† 0	—	5 15	5†45		—	6‡10			—	6 40	—	—		9† 0
Pwllheli (East)								—	—	—	—		—	—			—	—	—	7 10		—
Abererch								—	—	5 20½	—		—	—			—	6 45½	—	—		—
Penychain Halt								—	—	—	—						—	6 48½	—	—		—
Afon Wen								5†10	5 26	X5 30	5†55		6 20	X6 23	X6 20		6 51½	X6 54	7 20	7 25		9†10
Criccieth									—	X5 36			—	6‡29	6 26		—	7 0	7X36	7 52		
Black Rock Halt								Y	—	—	Y		—	—	—		—	7 4	—	—		
Wern Siding									—	—			—	—	—		—	—	—	—		
Stop Board									—	—			—	—	—		—	—	P	—		
PORTMADOC						—	3 30		5 44	X5 46			6‡37	X6‡39	6X34		7 11	X7 16	8X 9	—	8 15	
Granite Quarry Siding						—	—		—	—			—	—			—	—			8 22	
Minffordd						—	—		—	5 51			—	6‡44			—	7 21				
Stop Board						—	P		—	—			—	—			—	—	B			
Penrhyndeudraeth						3 49	X 4 5		5 54½	5 56			—	—			7 24½	X7 27	Passenger.		G	
Llandecwyn Halt						—	—		—	5 59½			—	—			—	7 30½	SO		L.E.	
Talsarnau						—	—		—	6 4			—	—			—	7 35	Runs July 30th to Sept. 3rd, inclusive.		FSO	
Tygwyn Halt						—	—		—	6 7			—	—			—	7 38				
Harlech						4 22	X●5 25		6 11½	X612½			6 58	6 59			7 43½	X745½				
Llandanwg Halt						—	—		—	6 18			—	—			—	7 51				
Llanbedr and Pensarn						—	—		—	6 20½			7	X5			—	7 53½				
Talwrn Bach Halt						—	—		—	6 23			—	—			—	7 56				
Dyffryn-on-Sea					V	5 45	X 5 55		—	6 28			—	—			— X	8 1	arr.	dep.	dep.	
Talybont Siding						—	—		—	—			—	—			—	—	p.m.	p.m.	p.m.	
Talybont Halt						—	—		—	6 32			—	—			—	8 5			N	
Llanaber Halt						—	—		—	6 36½			—	—			—	8 9½				
BARMOUTH	—	5 50		6 5	6‖11	6 12	6 18		6 40	X7 6		7 15	7 18X	7 25		7‖35	8 13	8U17	—	8 40	9‖40	
Barmouth Junction	5 55	5 56	6 10	6 11	6‖17	6 25	6 45		7 11	X7 12		7 21	7	30		7‖40	8U22	X8 23	8 45	8 46	9‖46	
Dolgelley Branch Loop									—	—			—	—					—	—		
Fairbourne									—	7 15½			—	—					—	8 50		
Llwyngwril									—	7 24			—	—					8 58	X9 0		
Llangelynin Halt									—	7 28			—	—					—	—		
Tonfanau									—	7 33			—	—					—	—		
Tonfanau Quarry Siding									—	—			—	—					—	—		
Towyn									—	7 38			7 53	7 54					—	9 13		
Aberdovey									—	7 45			8 0	8 1					—	9 20		
Penhelig Halt									—	7 48			—	—					—	—		
Abertafol Halt									—	7 53			—	—					—	—		
Gogarth Halt									—	7 59			—	—					—	—		
Dovey Junction									8 3	8‖18			8 15X	8 23					9	34		
MACHYNLLETH									8‖25	—			—	—					9 40	—		

A (Passenger to Wrexham): Runs August 6th, 13th, 20th and 27th only.
B (Auto to Dolgelley): Will not run on Saturdays, August 6th, 13th, 20th, and 27th.
B (Pass. and Mail): To Birkenhead.
A (Passenger to Aberystwyth): Runs on Wednesdays, July 20th to Sept. 7th, inclusive.

N—Engine of 6.35 p.m. ex Ruabon. R—Auto engine to Penmaenpool Shed.
U—Advertised 2 minutes earlier and must so run on Fridays and Saturdays. V—Engine of 3.5 p.m. (SX), 3.12 p.m. (SO) ex Ruabon.
Y—To work 6.20 p.m. Afon Wen to Pwllheli.
‡—Advertised Pwllheli depart 6.0 p.m., Afon Wen depart 6.20 p.m., Criccieth depart 6.26 p.m., Portmadoc arrive 6.34 p.m., depart 6.35 p.m., Minffordd depart 6.40 p.m.

No. 16 — 48

PWLLHELI, PORTMADOC, MACHYNLLETH and BARMOUTH.

Up Trains. Week Days. | Up Trains. Sundays.

STATIONS.	B Passenger. arr.	B dep.	B Auto to Dolgelley. dep.	B Pass. SX dep.	B Pass. SO dep.	K Goods. To Oswestry. arr.	K dep.	B Passenger. arr.	B dep.	B L.M.S. Pass. dep.	B Passenger. arr.	B dep.	B Passenger. arr.	B dep.	B Passenger. arr.	B dep.	K Goods. RR arr.	K dep.	B L.M.S. Pass. dep.	B Passenger and Mail. arr.	B dep.	A Passenger to Birkenhead. arr.	A dep.
	p.m.	p.m.	p.m.	p.m.	p.m.	a.m.	a.m.	a.m.	a.m.	p.m.	p.m.	p.m.	p.m.	p.m.	p.m.	p.m.	p.m.	p.m.	p.m.	p.m.	p.m.	p.m.	p.m.
PWLLHELI (Double Line)	—	8 10		9 10	9 36			—	11 15	12 15			—	3 15			—	—	5 0				
Pwllheli (East)	X	—		—	9X37½	Runs August 7th, 14th, 21st, and 28th only.		—	—	—			—	—			—	—	—				
Abererch	—	8 15½		9 15½	9 41½			—	1120½	1220½			—	3 20½			—	—	5 5½				
Penychain Halt	—	Z		—	—			—	1123½	—			—	3 23½			—	—	5 8½				
Afon Wen	8 22X	8 26		9§23	9Y48			1126½	X1132	12 26			3 26X	X3 30			—	4 30	5 11½				
Criccieth	—	8 32		9 29	9 54			—	11 38				3 35½	3 36½			4 43	X 5 17					
Black Rock Halt	—	—		9 32	9 57			—	—				—	3 40			—	—					
Wern Siding	—	—		—	—			—	—				—	—			—	—					
Stop Board	—	—		—	—			—	—				—	—			P	—					
PORTMADOC	8 40	8 55		9 40	10 5	—	6 0	11 46	—				3 47	—			5 35	—					
Granite Quarry Siding	—	—				—	—																
Minffordd	—	9 0				—	—																
Stop Board	—	—				P	—																
Penrhyndeudraeth	9 3½	X9 5				6 19	6 45																
Llandecwyn Halt	—	9 8½				—	—																
Talsarnau	—	9 13				C	R																
Tygwyn Halt	—	9 16				—	—																
Harlech	9 20½	X9 21				7 10	7 25				—	2 10			—	4 30							
Llandanwg Halt	—	9 27				—	—				—	2 16			—	4 36							
Llanbedr and Pensarn	—	9 30				7 35	7 45				—	2 20			4 38	4 39							
Talwrn Bach Halt	—	9 32½				—	—				—	2 23			—	4 42							
Dyffryn-on-Sea	9 37	X9 38				7 57	8 5				—	2 28			—	4 47							
Talybont Siding	—	—				—	—				—	—			—	—							
Talybont Halt	—	9 42				—	—				—	2 32			—	4 51							
Llanaber Halt	—	9 46			D	—	—				—	2 37			—	4 56							
BARMOUTH	9 50	9 55	10 15			8 20	8 35				2 40	2 50			5 0	—				—	5 55	—	7 15
Barmouth Junction	10 0	10 1	10 20		Empty C'ches	8 42	X 9 20				2 55	2 56								6 0	6 1	7 20	7 21
Dolgelley Branch Loop	—	—			SO	—	—				—	—								—	—		
Fairbourne	—	10 5				9 25	9 30				—	2 59								—	6 5		
Llwyngwril	—	10 14			dep.	9 42	9 54				3 9	X3 12								6 13	6 14		
Llangelynin Halt	—	—			p.m.	—	—				—	—								—	—		
Tonfanau	—	10 22				—	—				—	—								—	6 22		
Tonfanau Quarry Siding	—	—				—	—				—	—								—	—		
Towyn	—	10 28			10 47	10C15	X10T46				—	3 25								6 26½	6 27		
Aberdovey	— X	10 35			10 53	11 0	11 13				—	3 32								6 33	6 34		
Penhelig Halt	—	10 38			—	—	—				—	3 35								—	6 37		
Abertafol Halt	—	10 43			—	—	—				—	—								—	—		
Gogarth Halt	—	—			—	—	—				—	—								—	—		
Dovey Junction	10	52			11 8	—	—				3	48								6	50		
MACHYNLLETH	10 58	—			11 15	11 40	12 0				3 54	—								6 56	—		

Sundays B Passenger (11 15 a.m.), B L.M.S. Pass. (12 15 p.m.), B Passenger (3 15 p.m.) and B L.M.S. Pass. (5 0 p.m.): July 24th to September 11th, inclusive.
K Goods RR: RR August 7th, 14th, 21st, and 28th only.

Y—Arrives Afon Wen 9.47 p.m. Z—Calls to set down passengers on notice to guard. §—Arrives Afon Wen 9.21 p.m. Will cross 5.14 p.m. ex Birkenhead at Afon Wen on Fridays.

No. 16 — 49

BANKING AND SHUNTING ENGINES.

Engine No.	Starting Time	Finishing Time.	AUTHORISED HOURS. Mon.	Tues.	Wed.	Thrs.	Fri.	Sat.	Sun.	Total Hours per Week.	PARTICULARS OF WORK.
			PORTMADOC.								
1	7.30 a.m.	10.0 p.m.	14 30	14 30	14 30	14 30	14 30	14 30	—	87 0	Yard Shunting and Pilot Trips. To be ordered out specially when required to bank 4.30 a.m. Up Goods; bank 4.45 p.m. Down Goods when required.
2 3	10.15 a.m. 1.15 p.m.	11.45 a.m. 2.15 p.m.	2 30	2 30	2 30	2 30	2 30	2 30	—	15 0	Yard Shunting.
			PWLLHELI.								
1	7.30 a.m.	3. 0 p.m.	7 30	7 30	7 30	7 30	7 30	7 30	—	45 0	Yard Shunting.
2 3 4	7.30 a.m. 2. 0 p.m. 4. 0 p.m.	10.30 a.m. 3. 0 p.m. **SX** 10.30 p.m.	10 30	10 30	10 30	10 30	10 30	9 30	—	62 0	Passenger Train Shunting.
5	5. 0 p.m.	7. 0 p.m.	2 0	2 0	2 0	2 0	2 0	2 0	—	12 0	Yard Shunting.

Speed Restrictions

DOVEY JUNCTION AND PWLLHELI SECTION.

DOWN LINE.

The speed of any train between Dovey Junction and Pwllheli must not exceed 55 miles per hour, and must be further restricted to lower speeds as shewn below :—

Dovey Junction	Main Line	Coast Line	10
Dovey Junction	Station (Between 79 m. 0 ch.	Dovey River Bridge and 79 m. 20 ch.).	15
*Dovey Junction and Aberdovey	80 m. 40 ch. (Over Reverse	Aberdovey Station Curves).	25
Aberdovey	Any Train through Station		10
Aberdovey and Towyn	86 m. 15 ch.	86 m. 54 ch.	40
Towyn	Any Train through Station		10
Tonfanau and Llwyngwril	91 m. 59 ch.	92 m. 17 ch.	50
Llwyngwril	Any Train through Station		10
§*Friog Cutting	96 m. 26 ch.	97 m. 20 ch.	15
Barmouth Junction	Any Train through Station		15
Barmouth Junction and Barmouth Station	99 m. 40 ch.	100 m. 34 ch.	20
Barmouth Junction and Barmouth Station	Any train worked by engines other than "Yellow" and "Uncoloured" classes between 99	m. 41 ch. and 99 m. 79 ch.	10
Barmouth	Any Train through Station		15
Dyffryn-on-Sea	Any Train through Station		15
Llanbedr and Pensarn	Any Train through Station		10
§*Harlech Cliff	109 m. 20 ch.	110 m. 0 ch.	15
Harlech	Any Train through Station		15
Traeth Bach Bridge	115 m. 68 ch.	Penrhyndeudraeth Station	20
Penrhyndeudraeth	Any Train through Station		15
Minffordd	117 m. 20 ch.	Minffordd Station	20
Portmadoc	Any Train through Station		15
Wern Rock	121 m. 56 ch.	121 m. 70 ch.	20
Criccieth	Any Train through Station		10
Afon Wen Junction	Down Line	L.M.S. Line	10
Afon Wen Junction	Station	128 m. 60 ch.	30
Pwllheli East	Any Train entering Goods Yard		15
Pwllheli	Any Train entering Station		10

‡—See Note on page 61.

‡ This restriction applies only when the Signal Box is open. When it is closed trains may travel at ordinary speed in each direction.

§ Drivers to keep a sharp look-out for hand signals from Watchmen at Friog Cutting and Harlech Cliff.

*—Permanent Speed Restriction Indicator provided.

List of Signal Boxes

Distance between Signal Boxes. M.	C.	NAME OF BOX.	TIMES DURING WHICH BOXES ARE OPEN. Week Days. Opened. Mondays.	Other Days.	Closed at	Sundays. Opened at	Closed at	Whether provided with Switch.
		Whitchurch Junction to Aberystwyth—continued.						
5	24½	Dovey Junction	4.50 a.m.	4.50 a.m.	11. 0 p.m. **SX** 11.20 p.m. **SO**	**B**8. 0 a.m. 2.10 p.m. 6.15 p.m.	9. 0 a.m. 4.10 p.m. 7.15 p.m.	— — —
		DOVEY JUNCTION AND PWLLHELI.						
6	1	Aberdovey	5.20 a.m.	5.20 a.m.	10.55 p.m. **SX** 11.10 p.m. **SO**	8.20 a.m.**C** 2.30 p.m. 6.15 p.m.	9. 0 a.m. 3.50 p.m. 6.55 p.m.	— — —
3	49	Towyn	5.35 a.m.	5.35 a.m.	10.45 p.m. **SX** 11. 0 p.m. **SO**	8.40 a.m.**C** 2.45 p.m. 6. 5 p.m.	9.15 a.m. 3.35 p.m. 6.40 p.m.	— — —
6	44	Llwyngwril	5.50 a.m.	5.50 a.m.	10.30 p.m.	8.45 a.m.**C** 2.50 p.m. 5.45 p.m.	9.45 a.m. 3.30 p.m. 6.30 p.m.	— — —
3	68	Barmouth Junction	6.25 a.m.	6.25 a.m.	10.40 p.m.	8.55 a.m.**C** 2.45 p.m. 5.45 p.m.	9.55 a.m. 3.30 p.m. 7.40 p.m.	— — —
1	55	Barmouth South	6.30 a.m.	6.30 a.m.	10.25 p.m.	9. 0 a.m.**C**	10.15 a.m.	—
—	16	Barmouth North	6.10 a.m.	6.10 a.m.	10.20 p.m.	1. 0 p.m.**C**	7.25 p.m.	—
4	69	Dyffryn	5.55 a.m.	5.55 a.m.	9.55 p.m.	1.20 p.m.**C**	5. 5 p.m.	—
2	75	Llanbedr and Pensarn	5.25 a.m.	5.25 a.m.	10. 0 p.m.	1.30 p.m.**C**	4.50 p.m.	—
2	44	Harlech	4.45 a.m.	4.45 a.m.	10.15 p.m.	1.45 p.m.**C**	4.45 p.m.	—
5	39	Penrhyndeudraeth	4.20 a.m.	4.20 a.m.	10.25 p.m.	Clo	sed. **C**	—
3	27	Portmadoc	4.15 a.m.	4.15 a.m.	10.35 p.m.	**A C**		—
4	78	Criccieth	6.10 a.m.	6.10 a.m.	10.40 p.m.	**A**		—
3	28	Afon Wen	5.25 a.m.	5.25 a.m.	10.50 p.m.**B**	**A**		—
3	76	Pwllheli East	5.40 a.m.	5.40 a.m.	11. 0 p.m.	**A**		—
—	42	Pwllheli West	5.40 a.m.	5.40 a.m.	11. 0 p.m.	**A**		

A—Will open as required for services between Pwllheli and Portmadoc, July 24th to September 11th, inclusive.

B—Will close at 11.50 p.m. on Saturdays.

C—On Sundays August 7th, 14th, 21st and 28th will open for the passage of the Goods Trains.

Freight Workings

Details of the freight workings are not entirely clear and no official material has come to light about the allocation of freight guards workings. Nevertheless the pattern of freight traffic can be determined from the service timetables and working timetables available.

It would appear that through freight traffic was very limited, and that Barmouth or Barmouth Junction was regarded as a transfer point. Whilst this was logical for traffic coming off and going along the Dolgellau to Ruabon line, it seems less obvious why the division was made along the coast. The most obvious explanation appears to be that train crews and goods guards were limited by their working day as to how far they could work without relief being necessary, and with the GWR's preference to contain their brake vans to predetermined routes, these factors conspired to limit the freight workings.

Machynlleth goods guards worked as far north as Barmouth in the post war period, and some transfers were generally made at Barmouth Junction. Portmadoc guards had some trips as far as Barmouth Junction, usually with the early morning workings, and for a period, this was extended to Towyn where a change-over was effected. The first freight northwards pulled in to Barmouth yard where the working was handed over to Portmadoc men who had brought the first south bound freight as far as the Junction. Working EBV back to Barmouth they commenced the return journey as far as Portmadoc where the train pulled in to the yard. The working resumed after a couple of hours with a fresh guard to continue towards Pwllheli, allowing a spell of shunting and transferring wagons for the LM Region at Afonwen. All shunting and re-marshalling at Afonwen was done by the Western Region.

In the 1950's, most guards were kept on the same class of work, so freight guards rarely performed passenger train duties, although there were some mixed duties worked. Similarly there were some Barmouth porters who worked limited guards duties.

Portmadoc had by far the greatest volume of traffic to deal with after the war. Work emanated from Pwllheli Granite, situated between Minfordd and Portmadoc, and which saw regular movement of stone. Cookes Explosives at Penrhyndeudraeth also had daily movement of their products. The loaded vehicles were balanced by empty stock movements which provided much of the Portmadoc work for both the footplate and the guards. Pilot loco duties were necessary to work trains over the 'hump' between Portmadoc and Penrhyndeudraeth, a pilot engine being on call throughout the day – and much used of necessity. Shunting was undertaken for a couple of hours daily at Penrhyn, usually by one of the freight trains working north or south, depending on the time of day. Most traffic from the coast line worked south over Western Region lines, rather than working north to Afonwen and transferring to the LM Region. Similarly, the empty vans and wagons worked the same route, providing much work for the coast line. The remaining traffic was the usual coal plus agricultural produce and livestock. There was some limited slate traffic, but this was confined to the occasional single wagon load rather than regular consignments.

Distribution of brake vans is uncertain, but it is believed that there were four vans based at Portmadoc with six goods guards duties. At Pwllheli, there were three vans based at Pwllheli East with five duties, and it is believed that there were five vans based at Barmouth with eight duties.

DOVEY JUNCTION

Dovey Junction was for many years the interchange station with the Aberystwyth to Whitchurch/Shrewsbury lines.

Staff token points were as follows:- Machynlleth West to Dovey Junction; Dovey Junction to Glandyfi (or), Dovey Junction to Ynyslas when Glandyfi is switched out and Dovey Junction to Aberdovey. The station is situated at sea level and almost isolated with only a footpath for access.

The station was originally named Glandovey Junction but renamed Dovey Junction when Glandyfi Station, on the line to Aberystwyth, was opened. On occasions the high tides, which have been known to cover the tracks, have closed the station off and caused ballast to be washed away. Prior to the reorganisation of local government in 1974, the station was located at the point where 3 counties met. In fact, the platforms were in Montgomeryshire, the up distant signal in Cardiganshire and the station masters house in Merionethshire.

A description of signal box working at Dovey Junction can be found in 'Stopping Train Britain' by Alexander Frater, a book published by Hodder & Stoughton in 1983 and which features the line.

In steam days, many trains from Pwllheli terminated their journey at Dovey Junction, passengers being provided with connections to and from the Aberystwyth and Machynlleth lines. However, this necessitated light engine mileage and with the rationalization of the network most trains off the coast line now work to and from Machynlleth. The station was fully staffed until the above mentioned changes brought about the withdrawal of facilities and now trains do not stop unless it is necessary to cross with another or wait for a clear section ahead. Signalmen are the sole survivors and are housed in a flat roofed modern cabin of 65 levers, commissioned on 27th February 1959.

Run round loops enabled trains to cross and most used the platform lines wherever possible. In the summer seasons from 1953 until the Aberystwyth to Carmarthen closure of 1965, the 10.25a.m. Saturdays Only trains from Penychain to Carmarthen and Swansea, and the balancing working, the 10.10a.m. Swansea to Pwllheli, reversed here. For one season, Portmadoc men worked through to Aberystwyth where they changed footplates with the Down train from Swansea before working back to Dovey Junction, where they then reversed and continued to their destination. Manor class engines frequently worked this duty which was usually made up to eight vehicles. After one season the timings were altered slightly and Portmadoc men came off at Dovey Junction where they had a meal break and awaited the train from Swansea giving them about an hour and a half to wait.

1. Dovey Junction, 15th June 1965. Situated in the middle of marshes and away from any habitation, Dovey Junction served as an interchange for passengers, although since the introduction of the dmu, when locomotive shunting movements were eliminated, the station's importance has declined. In this view the Pwllheli portion of the Up 'Cambrian Coast Express' has drawn forward off the coast line towards Machynlleth in readiness to set back onto the stock from Aberystwyth, and prior to the departure of the combined portions for Shrewsbury and Paddington. BR Standard class 4 4-6-0 No. 75002 can be seen in the neck off the coast line platform. A dmu stands in the platform beyond the three coaches for Paddington. Note the low height four arm bracket signal, with the Down platform line arm off for the setting back movement. Also visible is the signal box with flat roof. *Photo: B. J. Ashworth*

2. Dovey Junction, June 1954. Outside framed 'Dukedog' 4-4-0 locomotive No. 9009 stands alongside the coast line platform with the 1.35p.m. for Pwllheli. A Machynlleth crew worked this duty, the driver on this occasion being Harry Benbow. This class of engine had numerous design variations with 9009 having a separate top feed. Note the platform canopy which was subsequently removed upon the rebuilding of the station in 1958/59. *Photo: W. G. Rear*

3. (Above) Dovey Junction, 11th June 1963. The layout at Dovey Junction was somewhat complicated and the fact that there were only two platform faces made station working difficult, particularly when it became necessary to cross trains as well as combine/detach portions. Passengers requiring information and assistance merely complicated matters, as time was usually short. On such an occasion, 2-6-4T No. 80101 stands at the coast line platform with a train from Pwllheli, with through coaches for Paddington. Meanwhile No. 7821 *Ditcheat Manor* waits alongside the Aberystwyth line platform with the Up 'Cambrian Coast Express'. It will detach and draw forward, whilst the 2-6-4T will also draw forward before setting back onto the Aberystwyth portion with the Pwllheli stock. After detaching, the tank engine will then draw clear allowing *Ditcheat Manor* to set back and couple onto the stock before pulling away to Machynlleth and Shrewsbury, where a replacement engine such as a 'Castle' or equivalent power will take over on the run to Paddington. *Photo: B. J. Ashworth*

4. (Centre) Dovey Junction, 5th August 1965. Both Pwllheli and Aberystwyth portions of the Up 'Cambrian Coast Express' meet alongside their respective platforms. Regular performer No. 75002 has travelled down the coast whilst the driver of 'Manor' No. 7812, without name and number plates, keeps an eye on proceedings. *Photo: J. M. Bentley*

5. Dovey Junction, 29th May 1964. A quiet moment looking east along the Aberystwyth line platform. All the station buildings had been replaced some six years previously. *Photo: C. L. Caddy*

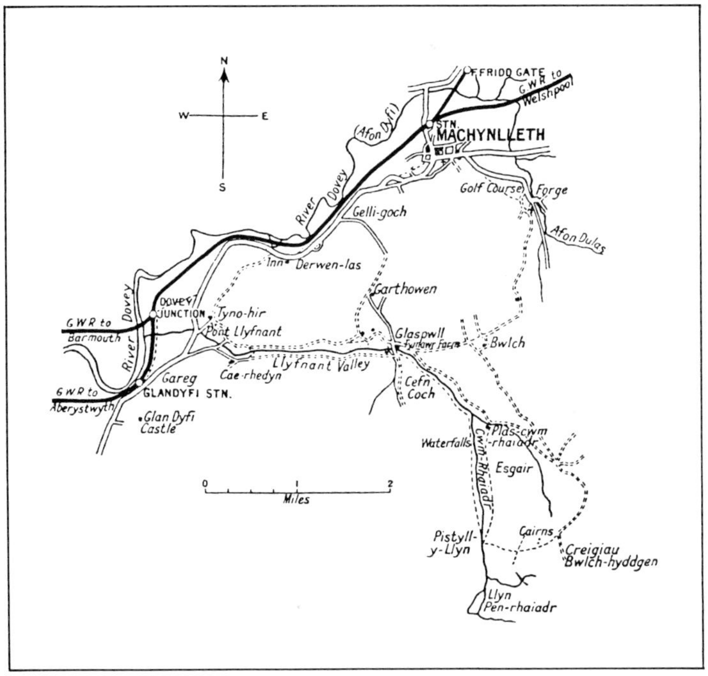

6. (Above-right) Dovey Junction, 28th July 1958. The all timber signal box containing a Duttons frame of 52 levers was closed on 27th February 1959. *Photo: J. J. Davis*

7. (Centre) Dovey Junction, 12th June 1973. Replacement signal box, again of wood construction situated on the platform with the coast line bearing off to the right. The frame was a British Railways Western Region standard type containing 65 levers. *Photo: C. L. Caddy*

8. (Below) Dovey Junction, 8th October 1958. Old and new on display following reconstruction of the station buildings in 1958. Utilitarian concrete structures are very much in evidence and one is left in no doubt as to where tickets can be obtained from! *Photo: British Rail*

GOGARTH HALT

An unstaffed halt located 1 mile 73 chains from Dovey Junction, running time for passenger trains from Dovey Junction was four minutes. The halt was erected during the 'twenties and came under the control of the Station-master at Dovey Junction. The platform was 99 feet long, built of timber and with a very basic shelter for passengers. The platform was located on the landward or Up side, facing the sea. One oil lamp, protective fencing and a name-board completed the furniture and access to the halt was from a footpath connecting with the road. There were no houses adjacent but several camping sites were nearby. In summer 1954, four Down and five Up trains called at the location with one other in each direction calling conditionally.

9. Gogarth Halt, 23rd May 1964. Gogarth Halt, opened by the Great Western Railway on 9th July 1923, was one of the first halts to be constructed after the amalgamation. This view, looking towards Dovey Junction, shows the basic construction of the halt. The corrugated iron shelter is typical GWR design.

Photo: C. L. Caddy

10. Gogarth Halt, 30th July 1963. Another view of Gogarth Halt, this time looking towards Aberdovey. The close proximity of sea shore, railway line and coast road can be clearly seen with the cliffs climbing away from the sea, indicating the problems encountered by the line's engineers during construction and subsequently by the maintenance department. An inspection trolley is to be seen beyond the platform. Progressive deterioration of the structure eventually resulted in the suspension of services from 14th May 1984, consent for complete closure taking effect from 28th September of that year.

Photo: H. B. Priestley

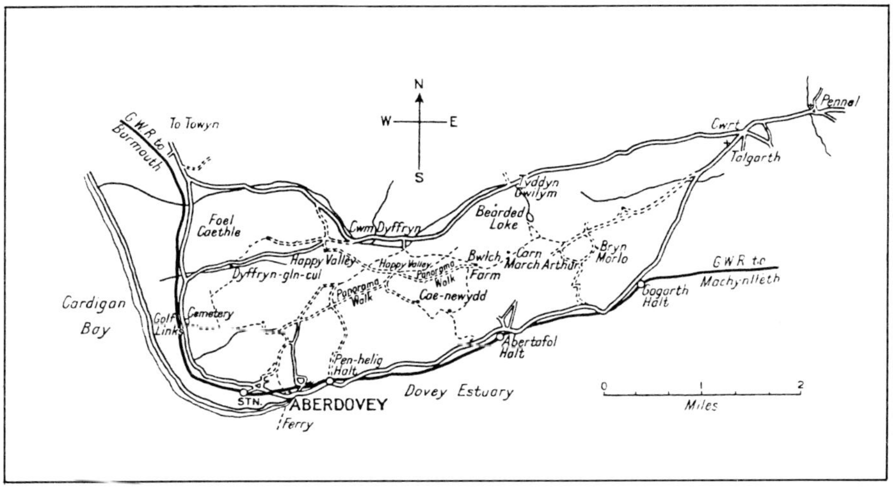

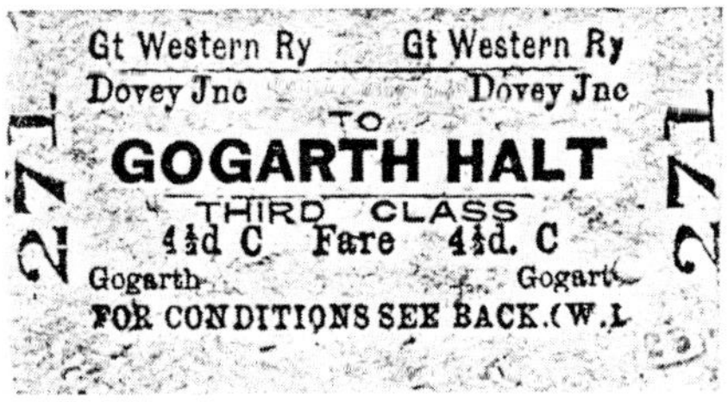

Gt Western Ry — Gt Western Ry
Dovey Jnc — Dovey Jnc
TO
GOGARTH HALT
THIRD CLASS
4½d C Fare 4½d. C
Gogarth — Gogarth
FOR CONDITIONS SEE BACK.(W.L.
271 — 271

11. Abertafol, c.1955. The crew of 'Dukedog' No. 9025 attracts the attention of a young lady walking along the adjacent road. The train is about to enter a short tunnel immediately east of Abertafol Halt before continuing along similarly constructed sea walls towards Penhelig and Aberdovey. These lightweight 4-4-0's with their low axle loading, were constructed especially for the Cambrian Coast line. The five coaches and van was a familiar formation, the class B headlamp indicating that the train was likely to call at every halt between the two extremities of the working. In all probability it is the 1.35p.m. from Dovey Junction to Pwllheli which frequently had a van attached as far as Barmouth where it would change to a Chester bound train via Dolgellau.

ABERTAFOL HALT

Another unstaffed halt on the long section between Dovey Junction and Aberdovey, construction being similar to Gogarth. The platform was again of all timber construction, but only 75 feet long, a situation which tested a driver's skill when required to halt with the guard's compartment level with the platform. The halt dates from 18th March 1935 when the GWR was exploiting every possible location to generate traffic. The shelter is even simpler than that seen previously but built of timber without windows. The statutory lamp, railings and name-board complete the fittings. Located on the landward or Up side the halt came under the control of Aberdovey Stationmaster.

Abertafol was 4 miles 11 chains from Dovey Junction, the journey taking ten minutes, with five trains in each direction calling at the halt in 1954. Recent economies have seen the halt taken out of use, closure taking place on 28th September 1985.

12. (Above-left) Abertafol, 11th July 1965. This view, looking towards Dovey Junction shows the platform construction. The track curves away across a low embankment, following the coastline before tunnelling for 219 yards under the headland. A few scattered houses near the main road could have provided some local traffic.
Photo: C. L. Caddy Collection

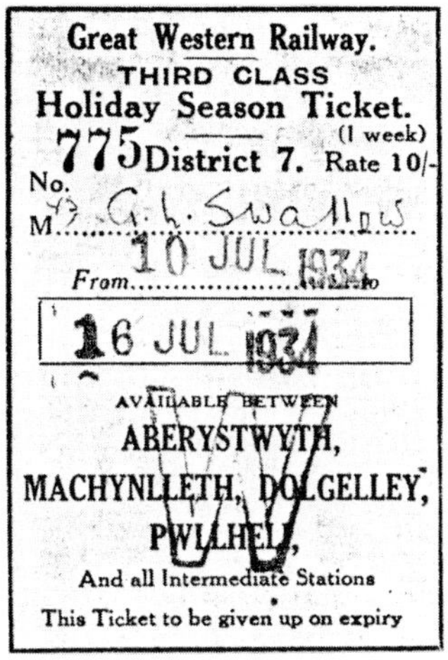
Great Western Railway.
THIRD CLASS
Holiday Season Ticket.
(1 week)
No. 775 District 7. Rate 10/-
M. G. L. Swallow
From 10 JUL 1934 to
16 JUL 1934
AVAILABLE BETWEEN
ABERYSTWYTH,
MACHYNLLETH, DOLGELLEY,
PWLLHELI,
And all Intermediate Stations
This Ticket to be given up on expiry

13. (Centre) Abertafol, 30th July 1963. A view of the platform looking towards Penhelig Halt and Aberdovey. The name-board identifies the location, which faces the Irish sea. Little comfort and not much shelter for passengers in this exposed and bleak location. *Photo: H. B. Priestley*

14. (Lower) Abertafol, 26th July 1955. A double-headed train from Dovey Junction to Pwllheli with Dukedog 4-4-0 No. 9009 as the pilot engine and Collett design 0-6-0 No. 2271 the train engine. It will be noticed that the track follows the coastline almost exactly, weaving in and out of every indentation in the landscape. Indeed, there is hardly a straight stretch of line between Dovey Junction and Aberdovey. On warm sunny days the view across the estuary is one of the finest anywhere, but in the winter months, with howling gale and spray blowing off the Irish sea, it was one of the less attractive features of the line and at times a railway engineer's nightmare.
Photo: Alan Donaldson

PENHELIG HALT

The third unstaffed halt between Dovey Junction and Aberdovey, opened on 8th May 1933, was significantly different from Gogarth and Abertafol in a number of ways.

Firstly, the halt was not totally unmanned, for in the high summer season, staff were on duty for part of the day. Secondly, the platform was much longer, being some two hundred feet in length and located on the Up side. The shelter was less basic in design, being of GWR 'Pagoda' design and with a small office located at the Aberdovey end. Finally, and probably most significantly, the station was located in the centre of a community, with about a hundred houses and dwellings nestling around the foot of the station. The line of track is not straight, but forms an elongated 'S'. The railway was forced to tunnel twice in the short space of a half a mile and the track being some fifty feet above sea level, causing the line to be built on an embankment. Also, the public road crossed under the road twice, as it followed the coastline, necessitating the line being raised sufficiently to clear reasonable loads using the road.

The estuary had been used for fishing and ship-building before the advent of the railway and there was a slip-way down to the water. The line provided a natural shelter behind which houses were constructed.

The platform, which is part solid part wooden construction, is located on the Up side of the line, with the aforementioned shelter and office, lamps, nameboard, a seat on the platform and protective fencing. Access to the station is by a footpath from the centre of the community below. At one time there was a siding at the Aberdovey end of the station but it saw little use and was subsequently removed when the GWR took over the Cambrian. There was a short length of track at right angles to the line, where a platelayers trolley was stabled and this can be seen on one of the photographs.

At the time of writing, Penhelig is still open, although like almost every other station between Dovey Junction and Pwllheli, it is unstaffed, the guard issuing and collecting the tickets.

In steam days, great accuracy was needed to draw up correctly at Penhelig, as sighting of the platform for drivers was somewhat difficult. However, every driver had his markers for every permutation of train length. The overhanging trees could make the rail slippery at times and when working trains over 200ft long, seven coaches or more, it became necessary to draw up twice. Difficult enough when one locomotive was at the head of the train, but as frequently happened, if a pilot engine was attached and with a ten coach train, this could mean that the leading engine was inside the tunnel-mouth with all the attendant difficulties that that presented. It is best left to the imagination as to conditions on the leading footplate, particularly if the coal was smoky or the fireman was having problems making steam. It was not unknown for personal scores between firemen to be settled in the Penhelig tunnels! Most trains stopped at Penhelig although one or two were listed as 'conditional'.

15. Penhelig, 26th May 1964. The platform at Penhelig, looking towards Dovey Junction. Penhelig tunnel is seen in the distance. At the end of the platform can be seen the girders and guard rails for the bridge over the road whilst the short length of track between the trees houses a permanent way trolley when not in use. The side windows of the shelter enable staff to see approaching Up trains from their office without going on to the platform in cold or inclement weather.
Photo: C. L. Caddy Collection

16. Penhelig, 5th October 1972. Looking towards Aberdovey, and showing the portal of Craig-y-Don tunnel, sometimes called Aberdovey tunnel. The bore of this tunnel was very close to the minimum limits and its 533 yards could ensure that crews on the train engine of a double headed working suffered extreme discomfort. The window in the Pagoda shelter lit the passengers waiting room although the interior was always gloomy. The platform was extended at some time and the wooden extension can be seen joining onto the original. *Photo: C. L. Caddy Collection*

17. Penhelig, 26th May 1964. A footpath over Craig-y-Don headland between Penhelig and Aberdovey provides us with this view of Penhelig station. Engine No. 82000 pulls into the platform with a Pwllheli bound train of ex GWR stock. The compact nature of the site is shown to its best advantage with houses backing right up to the line. The black and white building in the middle distance is an hotel, whilst the gardens of terraced houses can be seen beyond. The British Railways Standard class 3 2-6-2T, 82XXX class took over the duties of former GWR designs, particularly the passenger workings, between Dovey Junction and Pwllheli, but they in turn were replaced by the larger 80XXX class 4 design 2-6-4T's.
Photo: C. L. Caddy Collection

ABERDOVEY

Aberdovey is located 6 miles 64 chains from Dovey Junction, and was a staff token station with passing loop and sidings.

It started life as a temporary terminus for the line from Barmouth Junction until the extension to Dovey Junction was built. At one time it was proposed to construct a bridge over the Dovey Estuary linking directly with the Aberystwyth line.

There was a short branch line down to the harbour which remained in use until 1964. The connection to the harbour line was from the Down loop and vehicles were propelled down to the harbour with a brake van leading. The harbour line fell away steeply, and it was necessary to pin down brakes of descending trains to keep the train under control. Entrance to the harbour was through a set of gates, kept locked to trains, and the key was obtained from the signalbox before commencing the trip. The harbour line was difficult to work as sand frequently blew over the tracks and trains literally ground to a halt. Traffic to the harbour was quite heavy until after the first world war, when small coastal shipping diminished.

The station was fully staffed and under the control of a Stationmaster, remaining so until the loop was taken out of use. The station was enlarged in the early years of this century, sections of the building from the old Pwllheli station being re-erected at Aberdovey following the opening of Pwllheli (new) station. Subsequently the canopy was dismantled and taken to Llanuwchllyn, where it graces that station, now acting as terminus and headquarters of the Rheilfordd Llyn Tegid (Bala Lake Railway).

(continued over)

18. (Left-upper) Aberdovey, May 1963. Former GWR Collett 0-6-0 No. 2276 pulls out of the station with a train from Barmouth to Dovey Junction. The 1 in 70 grade is highlighted by the falling line to Aberdovey harbour in the foreground. The Cambrian had plans to develop the harbour into a port for Irish traffic and indeed some sailings took place. Alas, it was not to be, and by 1889 all attempts to establish a sea route had ceased. *Photo: C. L. Caddy*

19. (Left-centre) Aberdovey c.1950. A view of the main line looking towards Dovey Junction, taken from the end of the Up platform. The ramp for crossing the tracks is visible, as is the loading gauge in the loading dock. *Photo: J. H. Moss*

20. (Left-lower) Aberdovey, 26th May 1964. This view of Aberdovey station is taken from the loading dock and shows the main buildings on the Up platform as well as the Down side shelter and signal box beyond. The water column can just be seen on the Up platform, through the canopy, and beyond that can be seen a coach in the goods yard standing in front of the goods shed. The low height of the platforms was a feature of many stations on the coast line. *Photo: C. L. Caddy*

Aberdovey (continued)

The main buildings at Aberdovey were on the Up platform, with a semi open shelter on the Down platform. The signalbox was also located on the Down side and was a brick structure of less than standard height. It appears that the signalbox was erected before the Down platform was installed, and the lower part of the box was enclosed as the platform was built. When the passing loop was taken out of commission in the 1960's the Down loop was subsequently removed.

There was siding accommodation off the Up side at both ends of the station with the goods shed at the Towyn end. At the southern end was a single siding leading to a loading dock, whilst the siding accommodation at the northern end could store 50 wagons. The down side loop was 639 feet in length. There was also a water column located on the Up platform at the north end.

Running time to and from Dovey junction was 15 minutes for passenger trains and 25 minutes for class K freight trains.

There was a level crossing beyond Aberdovey, which had a gatekeeper, with the gates normally across the unclassified road.

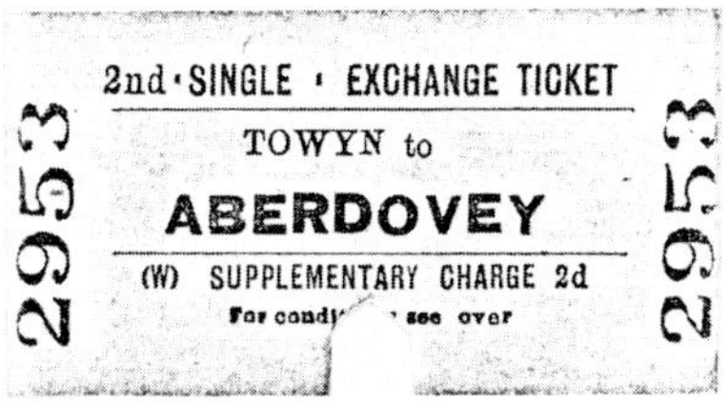

21. (Right-upper) Aberdovey, c.1950. Showing the station canopy from old Pwllheli station, and which is now at Llanuwchllyn. The buildings are of red brick with a steeply pitched roof. The main line climbs out of the station at 1 in 70, constrasting sharply with the harbour branch which falls away beyond the crossover to the single line. The lower quadrant Up starter stands alongside the unloading dock whilst the Down home is just visible protecting the Down loop. *Photo: J. H. Moss*

22. (Right-centre) Aberdovey, c.1950. A view north from beneath the down platform canopy. The goods yard, from which public facilities were withdrawn in 1964, can be seen beyond the Up platform. A water tank and column were also situated adjacent to the end of the platform. *Photo: J. H. Moss*

23. (Right-lower) Aberdovey, c.1950. The signal box at Aberdovey was of the small size, and comprised of a twenty lever frame. It was located on the Down platform and housed the train staff token equipment for the sections to Dovey Junction and to Towyn in the north. It also controlled access to the Harbour branch which extended off the Down loop. Reduction in line capacity resulted in closure of the box on 28th April 1968. *Photo: J. H. Moss*

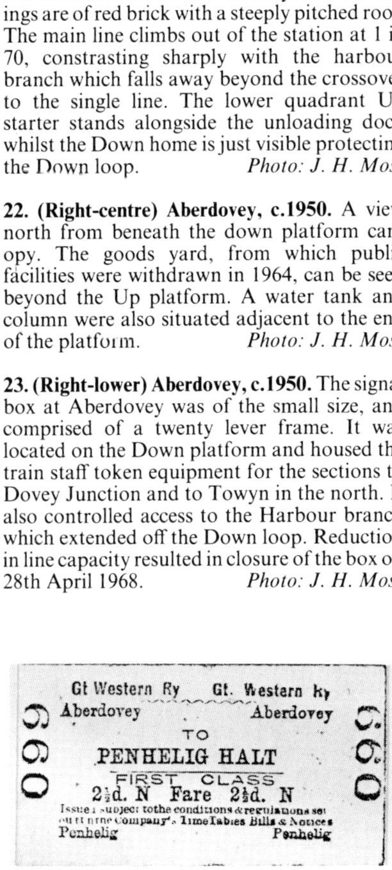

TOWYN

Towyn Station is 10 miles 39 chains from Dovey Junction, and another crossing point. Facilities at the station were generous and there was always plenty of freight activity in the yard. Some of the traffic was slate, generated in the hills inland and brought down to the coast by the Tal-y-llyn Railway which made an end on connection by the goods yard.

The main station buildings were again on the Up side although conditions on the Down side were somewhat spartan. The former were similar in appearance to those at Aberdovey although the canopy extended from the roof rather than rise and fall from the guttering as at Aberdovey. Passengers crossed the line by a footbridge of pre-cast concrete construction. The crossing loop on the Down side was 1218 feet in length, whilst the Up side extended 1389 feet.

The Down loop line extended beyond the entrance to the main line north of the station for about a hundred yards, terminating in a buffer stop. There was a refuge siding on the Down side which could accommodate 30 wagons plus the engine and brake van.

Beyond the station there was a level crossing which had a crossing keeper in residence, although the gates were normally across the road. Signals were also provided to protect trains. There was also a siding, known as Gas Works Siding on the outskirts of the town, on the Tonfanau (north) end, some way outside the station limits. Access was off the main line being made by Down direction trains which had to set back against a rising gradient, or by a local trip working from Towyn station. Access to the siding was by the train staff which locked the ground frame. In such cases, trains worked the trip with engine leading, and after detaching the brake van would be left on the main line, the transfer of wagons being made on re-coupling after the operation, the whole train would be propelled back to Towyn with the brake van leading. It was also permitted to propel trains for 2 miles 22 chains to Tonfanau Sidings or propel back to Towyn.

Apart from the slate traffic which was dealt with on the slate quay adjoining the Tal-y-llyn Railway

(continued on next page)

24. (Top) Towyn, 11th July 1964. A begrimed *Hook Norton Manor* No. 7823 runs past the slate wharf and the Tal-y-llyn terminus with a Down excursion. The 'Manor' class locomotives were infrequent visitors to this section of line mainly because the motive power departments could provide adequate power for the workings from the smaller tender and tank engines in their allocations. *Photo: C. L. Caddy*

25. (Centre) Towyn, 14th July 1964. A view that used to be familiar, but which has now gone for ever. The class K pick up goods, hauled by BR Standard class 3 tank No. 82020 working a northbound freight from Machynlleth to Barmouth Junction. The variety of wagons in the train should be noted, with an equal spread of mineral wagons and vans and with one tank wagon correctly located in the middle of the train as decreed by regulations governing unfitted loose coupled trains. Note the mobile crane on the transhipment siding, alongside the Tal-y-llyn Railway's Wharf terminus and the assortment of narrow gauge wagons just visible. *Photo: C. L. Caddy*

26. (Right-lower) Towyn, 17th July 1964. British Railways standard class 4 2-6-4T No. 80104 coasts into Towyn with a four coach local train. This was probably the the 1.35p.m. from Dovey Junction to Pwllheli for at this time, 80104 was based regularly at Pwllheli. The time can be accurately gauged by the appearance of the Tal-y-llyn Railway's locomotive *Douglas* at the western extremity of the narrow gauge line where it will no doubt refurbish its bunkers with coal. The Tal-y-llyn's lines on the tranship sidings had been cut back due no doubt to the decline in use. Note the jib of the rail mounted crane in the foreground. Just visible is one of the narrow gauge wagon turntables. *Photo: C. L. Caddy*

27. Towyn, c. 1954. Some three years after the takeover, ex Corris Railway 0-4-2, No. 4 *Edward Thomas* pauses at Wharf station, the Cambrian Coast line in the background. Much had been achieved in the short time since the organization had commenced operations but here things are very much unchanged. The siding to the right of the locomotive was later removed to make way for the narrow gauge museum building. *Photo: J. H. Moss*

27a. Towyn, c. 1954. Apart from a brief period in 1951, services continued between Towyn and Abergynolwyn. This view of the Wharf terminus shows No. 4 entering the station with a train of empty coaching stock prior to an outward journey. The locomotive, incidentally, was named after the former General Manager of the railway for over thirty years before the formation of the preservation society. *Photo: J. H. Moss*

28. Abergynolwyn, c. 1954. Arrival at the upper terminus of the line saw passengers deposited amidst some delightful scenery. Once again 0-4-2 saddle tank *"Edward Thomas"* duly obliges us with its presence. Over the next few years, the locomotive's appearance changed as first a running plate was added (1958) to enable the driver to fill the water tanks without standing on the motion and then a "Geisl" ejector was fitted (1959), in place of its chimney. *Photo: J. H. Moss*

Towyn (continued)

Wharf Station, there was a small goods yard on the Up side with a single road shed with internal crane for transhipping traffic from rail to road. The yard contained a coal siding and there was a small nest of cattle pens. The signal box was located on the Down side and contained the staff token equipment for the section south to Aberdovey and north to Llwyngwril.

In the late 1950's an excursion train ran from Rhyl to Towyn and return, worked throughout by Bangor men, using a BR Standard class 4 of the 75XXX class. The working was interesting, in that at first there was considerable light mileage. Once the stock was stored, the locomotive worked tender first back to Barmouth Junction where it turned on the triangle and returned to Towyn, again, tender first, where the crew cleaned the fire and attended to locomotive duties before taking up the return working. Driver Will Jones (Will Bach Bob Shunt) suggested to the LSM at Bangor that it would effect considerable savings if the locomotive could detach at Barmouth Junction on the outward trip, turn on the triangle and work the train tender first to Towyn, where they could dispose of the stock, perform locomotive duties and cut out all the light engine mileage. This was agreed, and the procedure adopted. It also gave the train crew another hour and a half to relax.

Passenger trains were allowed seven minutes to travel the 3 miles 55 chains from Aberdovey. There was sufficient freight traffic to justify an early morning working from Machynlleth to Towyn daily, running after the first Down freight train which worked to Barmouth Junction, and after the early morning Down passenger and Mails. It arrived at Towyn about ten past eight and shunted for an hour before running LE to Barmouth Junction (SX) where it worked an Up freight class K train back to Machynlleth, arriving there about 12.36p.m.

Tal-y-llyn Railway

Approaching Towyn from the south, we catch a glimpse of that pioneer of railway preservation, the Tal-y-llyn, whose Wharf Station can be seen to the right. The original terminus however was at Pentre when the railway first opened in 1866. Both passengers and goods traffic were carried, the latter chiefly being in the form of slate which was taken to the transhipment wharf alongside the Cambrian Coast line. Although the railway never closed, progressive deterioration of its equipment resulted in a takeover in 1950 by a group of men from Birmingham who initiated the Tal-y-llyn Railway Preservation Society. Total restoration since that time has seen the Tal-y-llyn go from strength to strength, culminating in the extension of services beyond Abergynolwyn to Nant Gwernol, a section formerly used only for mineral traffic.

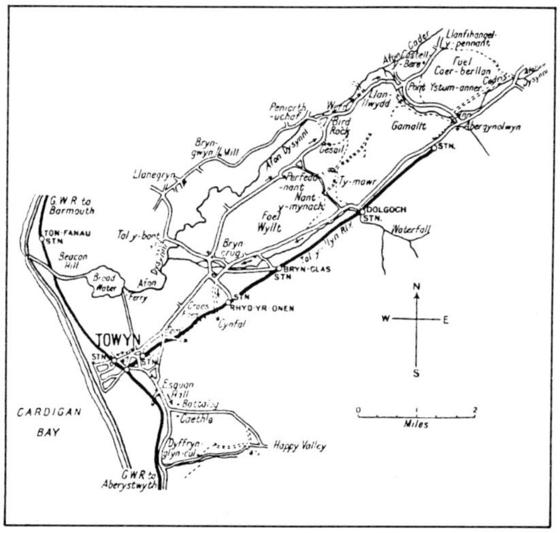

29. (Left) Towyn, c.1955. Taken from the Tal-y-llyn Railway wharf line, looking towards Towyn Station and beyond to Barmouth. Some shunting is in progress, with the brake van, two coaches and some mineral wagons in the Down refuge siding whilst the locomotive stands in the distance outside the signal box on the Down loop. The signal indicates that the main line is cleared for a setting back movement, but why remains a mystery. For many years, the Welsh Highland Railway 2-6-2T *Russell* stood behind the Narrow Gauge Museum under the shadow of the road retaining wall, exposed to the elements that blew off the sea, before being removed to the care of the Welsh Highland Railway Society for rebuilding. It finally came back to life in 1987, almost fifty years after it last steamed on WHR metals.

Photo: J. H. Moss

30. (Centre) Towyn, c.1950. Taken from the road over-bridge south of Towyn station looking towards Barmouth, this view gives a fine panorama over the complex. Notice the two arm signal, soon to be replaced by a modern tubular steel structure to be seen on other views elsewhere in this book. The down refuge siding is the extreme left hand track. The Down loop comes next, which constitutes the main line behind the camera. The Up loop follows with the crossover to gain the main line feeding in towards us. The goods yard is clearly identified with the coal siding on the extreme right in the yard. The cattle pens are in the middle of the yard, adjoining the shed road. Notice in particular the 'hump' in the centre of the station, where the lines pass over the road, beyond the footbridge.

Photo: J. H. Moss

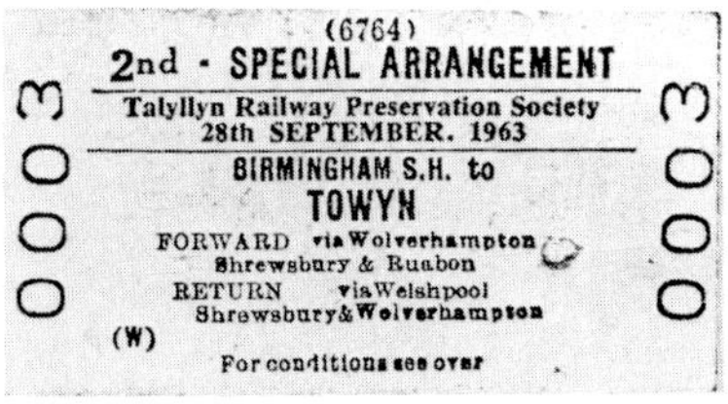

31. (Below) Towyn, 30th August 1959. British Railways Standard class 2 2-6-0 No. 78005 pulls a lightweight load out of Towyn with the 5.45p.m. local from Barmouth to Machynlleth. This journey was a Sundays only working and the GPO paid for the train – hence the van behind the tender. For many years it was the only train to run on the coast line on the sabbath although some excursions ran in the late 1930's from Birmingham or Shrewsbury to cater for the heathens! The train in fact worked north departing from Machynlleth at 9.00a.m. and arriving one hour later at Barmouth. The crew ran their engine around the stock in the excursion platform, thus avoiding the necessity of closing the level crossing gates at Barmouth. After performing their locomotive duties, they were left to amuse themselves until the evening before returning to Machynlleth where a connection was made with the mails from Aberystwyth to Shrewsbury. Again notice the two arm signal and the fact that the train is crossing from the Up side loop to the single line. In the distance, the signalman walks back to his box after exchanging tokens with the train crew on the platform. Twenty eight years later, an older sister to 78005 could be found in the same location, when 46443 from the Severn Valley Railway worked between Machynlleth and Barmouth twice daily, three days a week during the summer of 1987 with the Cardigan Bay Express.

Photo: M. Mensing

32. Towyn, 7th June 1963. Activity at Towyn. This beautiful view shows a BR Standard Class 4 2-6-4T No. 80104 on a Down class K freight of three mineral wagons and a brake van whilst a BR Standard class 3 2-6-2T No, 82033 waits in the Down side refuge siding attached to passenger coaching stock. In the goods shed the staff are dealing with a van. The signalbox with its token receiving equipment and the small platform where the signalman can deliver the token to Down trains by hand can clearly be seen as can the raised platform on the Up line alongside the goods shed, together with a standard lamp for illuminating the signalman. The ground shunting signals protecting the access to the Up line stand alongside the catch point whilst two ground disc signals face one another between the Up and Down lines, just beyond the walk way boards for the signalman. The Up starter is shown in all its detail. *Photo: B. J. Ashworth*

WESTERN REGION

Half-Day Excursion Bookings

EACH WEEKDAY

Commencing 12th June, 1961

UNTIL

Saturday, 9th September, 1961

TO

TOWYN

FROM	DEPART					RETURN FARE (Second Class)
		SO		SX	SO	s. d.
	p.m.	p.m.	p.m.	p.m.	p.m.	
BARMOUTH	12* 3	1 30	1 55	3 25	3 55	2/3
FAIRBOURNE	12*15	1 40	2 8	3 34	4 3	2/-
LLWYNGWRIL	12*28	1 48	2 17	3 42	4 14	1/6

SO—Saturdays Only **SX**—Saturdays Excepted
*—9 minutes later on Saturdays

Passengers may return by any Train Same Day affording a Service through to destination.

Children under Three years of age, Free; Three and under Fourteen years of age, Half-fare. (Fractions of 1d. charged as 1d.)

IT WOULD ASSIST THE RAILWAYS IN PROVIDING ADEQUATE ACCOMMODATION IF INTENDING PASSENGERS WOULD OBTAIN THEIR TICKETS IN ADVANCE

Further information will be supplied on application to the Stations or to Mr. O. VELTOM, District Traffic Superintendent, Shrewsbury (Telephone Shrewsbury 3614, Extension 42); or Mr. W. R. STEVENS, Divisional Traffic Manager, Cardiff.

Paddington Station, W.2.
May, 1961.

J. R. HAMMOND,
General Manager.

No. 66. R.P. 31 Printed in Great Britain by G. R. Griffith Ltd., Chester.

33. (Below) Towyn, 20th June 1970. Towyn signal box is a GWR standard timber design, replacing a Cambrian box on the same site. There were variations in roof design nevertheless and it was not uncommon to find a hip roof with the same structure underneath. This particular box housed a thirty-eight lever frame and controlled the signals manually, although the advent of radio controlled signalling will eliminate the need for visual signals. Nonetheless, Towyn is still in use as the remaining passing point between Dovey Junction and Barmouth. Notice in this view the lamp on the post to illuminate the token receiver arm which has a protective net should the train speed be on the high side, the net dampening the impact of the aluminium hoop. Notice also the finials, chimney and ventilator cowls on the roof, together with the array of telephone insulators behind the box. The walkway across the tracks are for the benefit of the signalmen when he needed to cross to the Up side. *Photo: C. L. Caddy*

34. Towyn, c.1955. Taken from the Up platform looking towards Aberdovey, this view well illustrates an item which was considered something of a rarity on the Cambrian Coast line, a footbridge and a pre-cast concrete one at that! The bridge along with many other items of platform furniture have long been removed to serve todays simplified requirements. Worthy of mention if only for the reason it has vanished is the Up starting signal of lower quadrant design. *Photo: J. H. Moss*

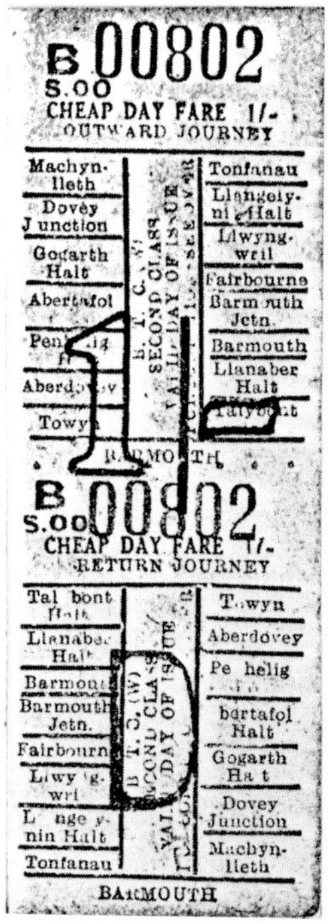

35. (Centre) Towyn, 23rd May 1964. Apart from new name-boards and totems the station changed but little over the years, and this scene could almost be considered as timeless. However, it will be noticed that beneath the large station name-board can be seen an additional sign, which reads "change for the Talyllyn Railway". In this view, the girders which supported the platforms and track over the road can clearly be seen and the track falls away from the bridge in both directions. Accommodation on the Down platform is very sparse, as this view shows, and it was the practice amongst those passengers who were 'in the know' to shelter on the Up side, and cross the line at the very last minute. There was little protection from the elements in the winter for passengers travelling north, and after all, there was always a fire in the Up side waiting room! *Photo: C. L. Caddy*

36. Towyn, c.1950. Another view of the 'hump' at Towyn, this time looking north towards Tonfanau. The exposed nature of the line is becoming more apparent and open cab locomotives could quickly lose their appeal when running tender first. Of particular interest are the canopy support columns which could have severely inhibited passengers hurriedly leaving an arriving train. *Photo: J. H. Moss*

37. Towyn, 17th June 1962. The line between Towyn and Tonfanau follows the sea shore and in an effort to combat erosion large boulders were placed alongside the track as a form of protection. In this photograph the stone has been emptied from a wagon and placed into its prescribed position at the track side. The empty wagon is then lifted bodily by the crane and swung through 180 degrees and placed back on the track where it is moved out to join other empty vehicles. Meanwhile another wagon with boulders is moved up towards the crane. By adopting this method, time is saved through not having to shunt empty wagons with a locomotive and thus avoid time consuming movements. The scheduling of such an exercise ensured that every movement was planned with stock being positioned by Saturday night, to enable the Civil Engineering department to obtain possession early the following morning. *Photo: B. J. Ashworth*

38. (Below) Tonfanau, 5th October 1972. Although taken after steam had long since disappeared (albeit temporarily) from the Cambrian Coast line, Tonfanau Station had hardly changed and this view could have been taken at any time from the 1940's to late 1970's. Perhaps the buildings are shabbier, and the fences lacking in paint. What appears to be a sheep grid straddles the line, and the station name sign has progressed from a brown totem to a black and white rectangular strip, but otherwise time is standing still. *Photo: C. L. Caddy*

TONFANAU

Tonfanau was an intermediate station on the 4 mile 47 chain section from Towyn to Llwyngwril and 12 miles 71 chains from Dovey Junction, passenger trains taking 42 minutes to reach the halt.

There had been a large Army camp built on both sides of the line, which had its own complex of sidings and platforms subject to military control.

Considerable mineral traffic emanated from the quarry on the Towyn side of the station and it was the practice for trains to work out from Towyn to shunt the quarry sidings, and subsequently propel the train back to Towyn as permitted. In addition, the line fell steeply towards Tonfanau at gradients varying from 1 in 175 to 1 in 75 which necessitated freight trains exercising caution whilst shunting the quarry siding.

The station buildings are located on the Down side of the line and are fairly simple in construction. There was never much civilian traffic outside the holiday period and basic shelters sufficed although the station had staff on duty during the day throughout the year. The location of the station was nevertheless somewhat exposed and the platform could be a bleak place, especially in winter.

A level crossing was located at the Towyn end of the station, the gates coming under the control of the station staff and were normally left against the railway and open to the road traffic. A train operated treadle is located one thousand yards away.

The station is currently open to traffic although it has been unstaffed for many years. The military and mineral sidings have been removed.

39. Tonfanau, circa 1960. Tonfanau Station served a large army camp, which had its own sidings and platforms, the only trace visible in this photograph being the large building behind the simple shelter on this platform. BR Standard class 3 2-6-2T No. 82005 draws into the station with a three coach working for Pwllheli. Apart from the station staff on the platform, there appear to be no passengers.
Lens of Sutton

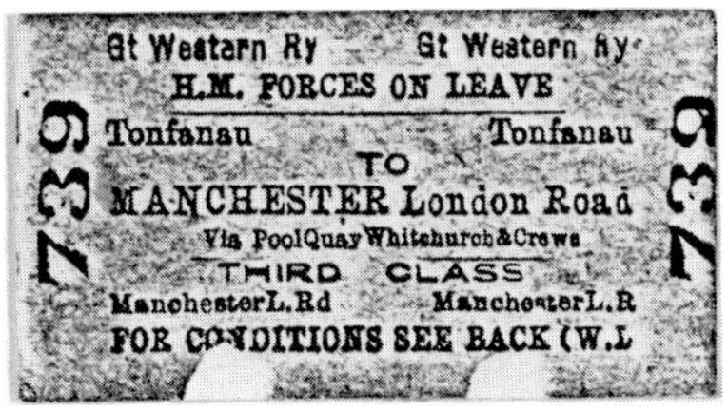

LLANGELYNIN

Llangelynin Halt is 15 miles 6 chains from Dovey Junction on the single line section from Towyn to Llwyngwril. It was and still is an unstaffed halt, opened by the GWR on 7th July 1930. As is usual with halts erected at this time, the layout is basic. The wooden platform edge with earth infill is 75 feet long. Passenger accommodation is a simple sloping roofed open fronted shed built of corrugated iron sheeting. The remaining furniture comprised of the usual name-board and two oil lamps. The platform was located on the Up side of the line, and sheltered by an embankment, access being by a footpath from the road which ran about ten feet higher. The summit of the line was nearby and sighting distance to the platform was short in both directions, another instance where drivers of steam hauled trains had to have markers to enable them to stop with the guard's compartment at the platform face. It must have been somewhat difficult for trains of eight coaches to receive the guard's flag and/or lamp signal before restarting.

There have been few apparent changes over the years, the only noticeable difference being the name-board with its universal black letters on white background, British Rail's corporate image offering little to improve the appearance of this lonely spot.

40. (Above) Llangelynin Halt, 28th May 1964. The lonely location of Llangelynin Halt is emphasised by this view of the platform, with its shelter, two lamp pillars and solitary name-board. The platform edge is formed out of old wooden sleepers, running along the edge, whilst the old GWR style name board still retains its large individual letters screwed to a brown painted background board. The sharp curve looking towards Towyn seems to draw attention to the isolated situation of the halt.
Photo: C. L. Caddy Collection

41. Llangelynin Halt, 2nd October 1971. Llangelynin (the 'Halt' has now been dropped from the title) stands unchanged apart from the new name-board. The track looking towards Barmouth falls away and one can sense the relief a driver and fireman might feel as they breast the summit, gaining some respite from the change in gradient as they struggle towards Dovey Junction with an unfitted freight.
Photo: C. L. Caddy

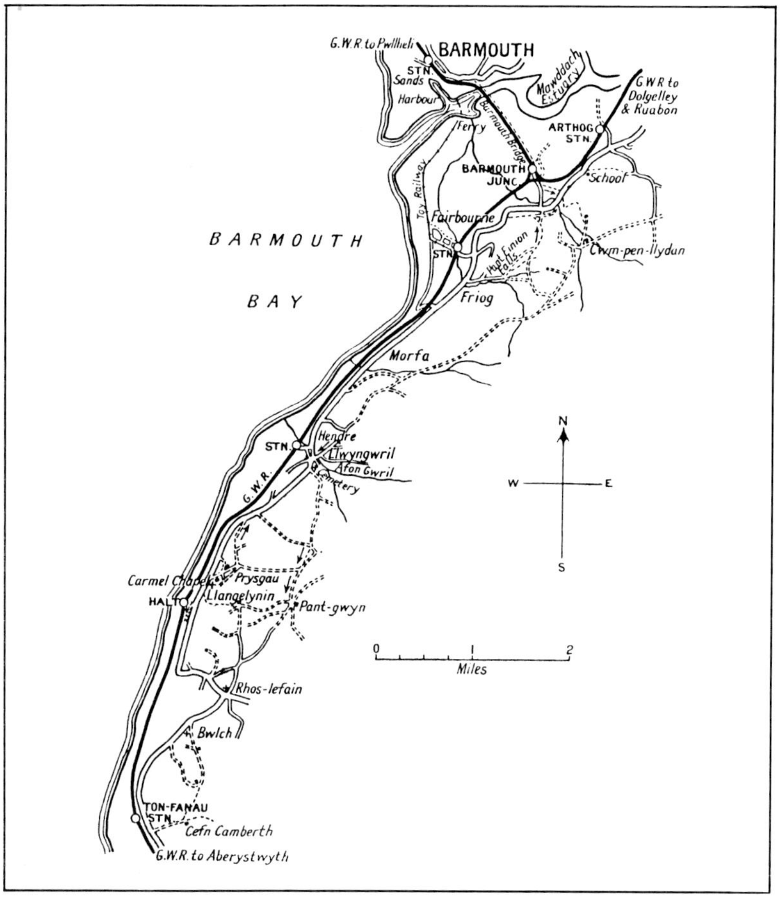

LLWYNGWRIL

Seventeen miles and 1 chain from Dovey Junction is Llwyngwril, a passing point on the single line to Barmouth. The line has seen some severe gradient changes in the six miles or so since leaving Towyn, at one point as steep as 1 in 75. Passenger trains generally have taken 43 minutes to travel this far, whilst the early morning Down freight train has taken 65 minutes to travel from Machynlleth, pausing only at Towyn to take on water.

The station consisted of the usual passing loop, both Up and Down lines being 729 feet in length. The main buildings are located on the very short Up platform, in fact three coaches was the maximum train length that could be accommodated without having to draw up a second time. The Down platform was slightly longer. The large water tank which dominated the sky line was located at the Barmouth end of the station and fed the water column at the Towyn end of the Up platform and at the Barmouth end of the Down platform. There was a single siding also on the Up side, which was little used and after the second world war was used to hold a camping coach. Passengers crossed the line by a walk-way at the Barmouth end of the platform ramps. The Down platform had a very basic shelter, somewhat larger than at Llangelynin, but with windows to aid visibility. The platform height was lower than usual and steps were provided for the convenience of the passengers boarding and alighting trains. On the Down side was a very small Cambrian Railways design signal cabin, which housed a thirteen lever frame which survived until the passing loop was discontinued. The cabin contained the electric tablet equipment controlling the single lines to Barmouth Junction in the north and Towyn in the south.

Llwyngwril Station remains open with a remaining single line serving the Up platform, a layout which precludes of course the passing of trains at this point.

42. Llwyngwril, 23rd May 1964. As previously mentioned, the Down platform was much longer than that on the Up side and evidenced by this view taken from near the signal cabin. The main building looks clean and tidy and there is an air of care and devotion about the place. Even the water column paint looks fresh. The lamp supports on the Down platform however are shorn of their fittings, just the hooks remaining. Beyond the Down starter can be seen an un-gated crossing. Whilst a member of the station staff walks away from the Up platform – is he the signalman?

Photo: C. L. Caddy

43. Llwyngwril, circa 1963. BR Standard class 3 2-6-2T No. 82005 heads a Pwllheli to Dovey Junction train. Note that the first coach is off the platform, whilst the guard can be seen in the distance boarding his van. The signalman waits in the siding line whilst the train draws away before crossing to his cabin and informing Towyn that the passenger train is on its way. Note the extension to the signal box, with the overhanging extension supported by brackets secured to the brickwork. The long chimney on the cabin roof has twin vents. The shrubs on the down platform indicate that someone on the station staff looks after the garden.
Photo: J. W. T. House, C. L. Caddy Collection

44. Llwyngwril, 23rd May 1972. This small signal cabin at Llwyngwril was of Cambrian Railways origin, housing thirteen levers. It was located at the southern end of the Down platform, and contained the train staff tablet equipment. Note the two bells and limited forward visibility, although the end windows were adequate to sight trains approaching. The modern pre-cast coal bunker and shed alongside contrasts with the cabin and even the nameplate is smaller than usual. Nevertheless this design of cabin was not unique, and similar models could be found on the Moat Lane to Brecon line, also of Cambrian Railways origin.
Photo: C. L. Caddy

45. (Below) Llwyngwril, 23rd July 1951. Llwyngwril Station looking towards Towyn showing the simplicity of the layout although the main building is quite substantial. Once again 'Dukedog' No. 9017, member of a class which was the mainstay of Cambrian main line services, heads a Pwllheli train. In the siding to the left stands a camping coach, its safety from other rail vehicles being effected by removal of rails in front. The only limited goods facilities were withdrawn on 4th May 1964. *Photo: R. C. Riley*

FRIOG

No survey of the Cambrian Coast line can be regarded as complete without mentioning Friog, where the line clings to the cliff-side, its path hewn from solid rock, and where on two occasions, trains have come to grief with tragic results.

The line reaches a height of one hundred feet above sea level, with the road running a similar height again above the line. The climb to Friog was arduous and there was always caution by train crews whilst working this section. The rocks above the line were liable to erosion by the elements and when some mining activity took place in 1870 there was cause for concern that the workings of the mining company might disturb the strata. It was suggested by the Railway Inspectorate that the problem could be avoided by tunnelling a deviation, but the Cambrian Railway was in no position to respond positively. On 1st January 1883 a landslide blocked the line and the evening train from Machynlleth ran into the obstruction, derailing the locomotive which fell over the edge onto the rocks below, killing the train crew.

Some fifty years later, on 4th March 1933, an accident occured at the same location when the 6.10a.m. morning mail from Machynlleth to Pwllheli struck debris near the same spot, sending the locomotive to the shore below again killing its train crew in the process. In both accidents, the engines plunged to the sea, but the coaches, although derailed, remained on the trackbed. Consequently a reinforced concrete avalanche shelter was constructed to eliminate the risk of further accidents.

46. (Top) Friog, 28th July 1951. A view looking towards Barmouth showing the extent of the protective works largely carried out by the GWR. Immediately to the right of the oncoming train is the section of cliff face strengthened with masonry and tied with rails. Next to this, a slightly more sophisticated and expensive method extending further up the cliff face shows the works strengthened by relieving arches. The third and final method of protection involved the construction of an avalanche shelter consisting of a series of heavily reinforced concrete arches supported on one side, the cliff edge, by concrete piers, and the other, a rock face into which the arches are deeply set. Running the entire 60 yard length is a twelve inch reinforced concrete roof pitched at an angle that is intended to direct any falling debris safely to the shore below.
Photo: R. C. Riley

47. (Centre) Friog, 28th July 1951. The avalanche shelter section viewed from the southern end. To the left of the structure in the centre of the picture is the sea wall at Fairbourne. *Photo: R. C. Riley*

48. (Right) Friog, 28th July 1951. Approximately one hundred feet above the railway is the coast road from which point this view of an unidentified 'Dukedog' is taken as it descends to Fairbourne. A constant bane on the budget of the railway engineer, the cliffs at this point still require constant surveillance and maintenance.
Photo: R. C. Riley

49. Friog, c.1950's. This postcard view date stamped 1st July 1958 again shows the effects of erosion to the cliffs at Friog. The road has been diverted following a further bout of protective works. The cluster of buildings to the left centre of the picture includes Fairbourne Station, giving an indication of the fall in levels from this point. *Photo: Mrs V. Palmer Collection.*

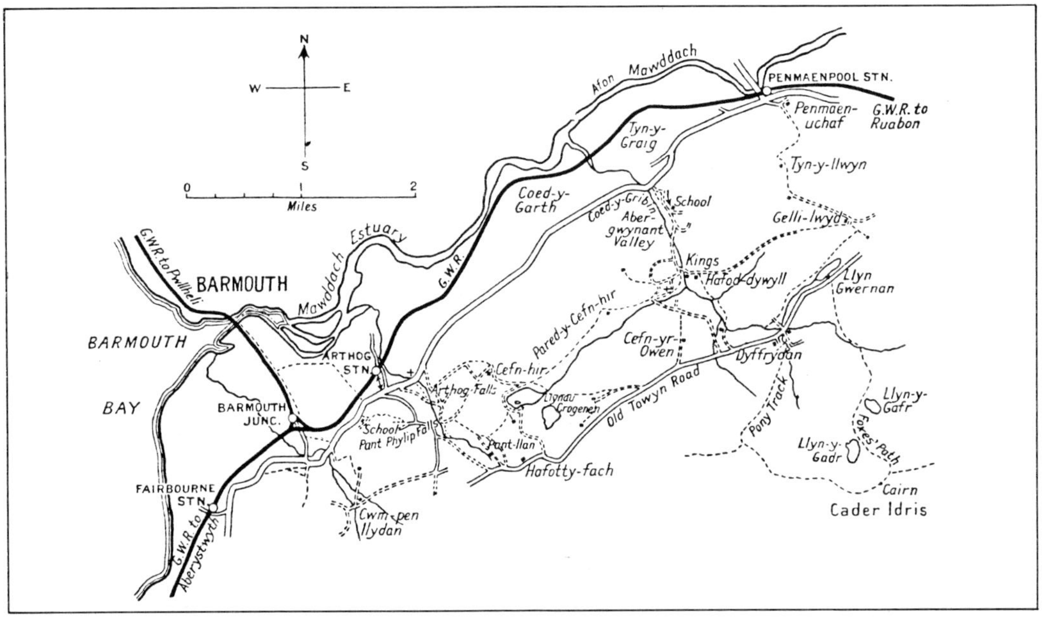

FAIRBOURNE

Fairbourne Station is at the foot of the falling grade from Friog, the line descending for about a mile at gradients varying from 1 in 76 to 1 in 55 at its steepest point. The line runs over a level crossing before the station is reached, the platform being on the Down side. It was not a staff token station but a halt on the section from Llwyngwril to Barmouth Junction. The crossing gates were under the control of the Stationmaster or his staff, signals controlling train movements and interlocked with the crossing gates, which were normally left across the rails and open to the road users.

Just before and to the left of the level crossing can be seen the southern terminus of the Fairbourne Railway, which runs amongst the sand dunes to a point opposite Barmouth, a ferry linking the two points. Latterly, the Fairbourne Railway has had a change of owner and in a radical move completely rearranged the track from fifteen to twelve and a quarter inches. The locomotives and rolling stock are now of narrow gauge outline.

The station at Fairbourne was staffed until the general removal of personnel in the early 1970's. Buildings are minimal with a canopied shelter offering limited protection to passengers. Together with the Stationmaster's house on the platform, there are one or two other buildings, their construction being a mixture of wood and corrugated iron sheeting.

A single siding, situated on the Down side and facing Barmouth Junction saw little traffic and after the last war was used to house a camping coach. Access was via a ground frame alongside the point, which in turn was locked by the train staff. The gradient of 1 in 76 resumes after leaving the station which is 19 miles 54 chains from Dovey Junction, and 72 chains from Barmouth Junction. Passenger trains took fifty five minutes to cover the journey. Goods facilities, such as they were, were withdrawn from 4th May 1964.

50. (Left) Fairbourne, c.1960. Fairbourne Station platform, looking towards Friog and Towyn. The track can be seen climbing away from the station to gain height. The somewhat austere look of the Stationmaster's house is probably no reflection on the comfort level, for conditions were usually very snug. In the foreground can be seen a trolley loaded with milk bottles in their crates, a traffic which was commonplace over the whole network, but has since disappeared. Beyond the level crossing can be seen the gable end of the Fairbourne Railway terminal building, an indication of the difference in height that the main line has scaled. The signals protecting the crossing are 'off' and the gates are across the roads, indicating that a train is imminent, a fact reinforced by the presence of a member of the station staff visible by the gates, and ready to re-open them to road traffic once the train has cleared the crossing.

Photo: Lens of Sutton

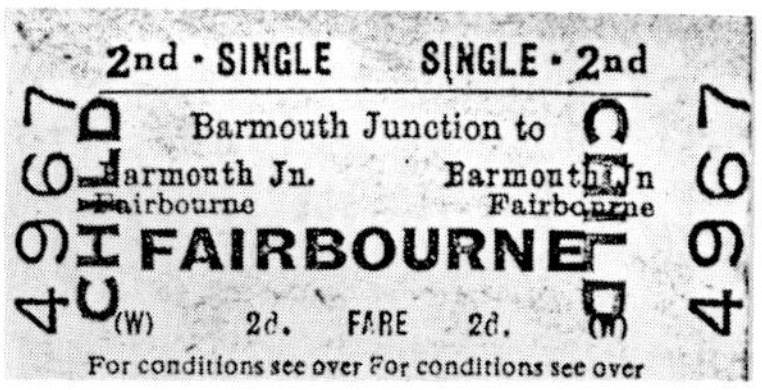
2nd · SINGLE SINGLE · 2nd
4967 CHILD Barmouth Junction to CHILD 4967
Barmouth Jn. Barmouth Jn
Fairbourne Fairbourne
FAIRBOURNE
(W) 2d. FARE 2d. (W)
For conditions see over For conditions see over

51. Fairbourne, 23rd September 1961. View at the terminus of the Fairbourne Railway with Mr J. C. Wilkins, director of the railway, preparing the 15 inch gauge locomotive *Count Louis* for a working to Penrhyn Point. The corrugated building in the middle distance is the engine shed whilst British Railways' station can be seen above the two carriages.
Photo: Norman Jones

52. Fairbourne, 23rd September 1961. Having reached Penrhyn Point the engine would run round its train for its journey back, tender first. This view across the bay to Barmouth shows the 4-4-2 *Count Louis* preparing for the return trip. Two passengers make their way across the sand to the waiting ferry boat for the short sail to Barmouth, seen here in the background.
Photo: Norman Jones

53. (Below) Fairbourne, 23rd September 1961. A train from Fairbourne to Penrhyn Point running beside the sea wall. The locomotive *Count Louis* with its 'Atlantic' 4-4-2 wheel arrangement, was a Bassett Lowke improved version of the famous Little Giant class. It was completely rebuilt in 1947, the makers designation being Series III Class 30. Originally built in 1924, *Count Louis* was built for the racing driver, Count Louis Zbrowski, who had his own private railway. Sadly, he was killed in a motor racing accident.
Photo: Norman Jones.

54. (Right) Fairbourne, 23rd May 1964. Fairbourne has been and still is a popular holiday resort, both the GWR and British Railways playing their part in promoting its appeal. The railways were shrewd enough to know that a camping coach ensured that anyone hiring such accommodation for a holiday had to travel to it by train with a minimum number of tickets. Nevertheless a coach would not be installed at a location where it was thought likely to be difficult to keep fully booked throughout the season. Alas, this type of holiday was on the decline, and there were signs of modern affluence. Note also the overhead wire warning flashes on the coach ends, another latter day requirement as the railway entered into its modernisation programme. *Photo: C. L. Caddy*

55. (Centre) Fairbourne, 4th October 1962. The station at Fairbourne was originally opened on 6th June 1899 by the Cambrian Railways on the site of the temporary Barmouth Ferry station which had closed on 3rd June 1867 when Barmouth Bridge was opened. The company however, introduced a round trip facility, using the standard gauge from Barmouth to Fairbourne, the Fairbourne Railway to Penrhyn Point and a ferry boat across the waters of the Mawddach estuary. In this view, things have changed little. There are still milk bottles to be carried to somewhere, on this occasion a sack truck, but the camping coach is still in the siding. In the distance, just visible through the fencing past the far platform ramp can be seen the spans of Barmouth bridge.

Photo: The late J. K. Nelson

56. (Below) Fairbourne, 25th May 1954. Collett 0-6-0 No. 3201 with an 89C (Machynlleth) shed plate coasts into Fairbourne station on the short journey from Barmouth Junction. The smoke from the chimney indicates that the fireman has built up his fire in readiness for the climb to Friog and Llwyngwril. The points in the foreground lead from the single siding mentioned previously, site of the camping coach. The tender attached to 3201 is of a type more familiar to the Dean 'goods' locomotives. *Photo: L. Marshall*

57. Barmouth Junction, circa 1910. A view overlooking the estuary showing Barmouth Junction Station, with the Llangollen line curving in from the right and meeting the coast line which runs across the centre foreground of the picture. The bridge over the estuary extends away in the distance. Notice also the cottages that once stood in the triangle, the goods wagons in the siding and the line of wagons in the foreground. Cambrian Railways vehicles are conspicuous by their absence.
Photo: Mrs V. Palmer Collection

BARMOUTH JUNCTION

Barmouth Junction was 20 miles 62 chains from Dovey Junction and 52 miles 17 chains from Llangollen Line Junction signal box near Ruabon. Running time for passenger trains from Dovey Junction was fifty eight minutes.

The station was located at the boundary of the Oswestry Division, which controlled the Cambrian Coast line, and the Chester Division. The station itself was under Oswestry supervision. Originally named Barmouth Ferry as a replacement of the temporary station on the Fairbourne site, it had become known as Barmouth Junction by 1871, retaining that title for the best part of a century until 13th June 1960 when British Railways renamed the station Morfa Mawddach.

It was the creation of the Dolgellau branch that established the junction, or more precisely when the through route from Ruabon became effective enabling passengers and freight traffic from the Dee Valley travelling towards Aberystwyth to avoid travelling into Barmouth and retracing their steps.

The junction was triangular which enabled locomotives to turn without the need to continue into Barmouth to use the turntable. The third leg of the triangle from the coast line to the Dolgellau line was restricted to light engines. With some freight traffic terminating and starting from this point, it was the practice for some locomotive duties to turn here prior to working their return journey. For example, Portmadoc men working the early morning goods turned on the triangle, as did the Machynlleth men working northwards with their early morning goods. However, there were a couple of sidings but no goods yard at Barmouth Junction, and freight traffic scheduled to terminate here was in fact worked forward by the respective divisions. The same was true for those freight workings working north or south along the coast line, and which used Barmouth Junction as a transfer point. Freight trains which required sorting, however, were dealt with at Barmouth. However, according to the Appendix for the service timetables covering No. 16 section, dated March 1943, there was a small goods warehouse located on the down platform.

There were platform faces for the Up and Down coast lines, and similarly for the Llangollen line. The Down line from Llangollen and the Up coast line shared the island platform. Passenger access to the outer platforms was via walk-ways at the end of the ramps at the Barmouth end. The one signalbox, which measured 31′ 0″ by 10′ 0″, was located on the up side at the junction of the lines and was a standard GWR design constructed of wood, the frame having 38 levers. A walk-way from the signalbox across the tracks led to the centre of the coast line, where the signalmen would exchange single line tokens with the train crew of Barmouth bound trains.

The points controlling the third leg of the triangle were hand operated. On the Dolgellau line, the point was located on a head-shunt to the down line, some distance beyond the crossover from the single line, which was a continuation of the Up line. This then connected with the Up line head-shunt of the coast line, some distance beyond the crossover at the commencement of the single line.

The working of the swing bridge section was set down in the sectional appendix and the operating mechanism was interlocked with the tablet instruments between Barmouth, the Bridge Cabin and Barmouth Junction. If there was a token withdrawn from the instrument the operating mechanism could not be unlocked. If the bridge was required to be opened, the keeper at the Bridge Cabin would advise the boxes on either side, who would then depress the tablet instrument plunger enabling the Bridge Cabin keeper to withdraw two keys. This in effect broke the circuit. The keys unlocked the bridge bolt levers and the bridge could be opened. The procedure was reversed after the bridge was closed. Once the keys were replaced in the token instrument, the circuit was restored and the section was available for the passage of trains – but not before the signalman at Barmouth Junction had tested the circuit by ringing sixteen consecutive beats to Barmouth South signal cabin.

Although there was no proper road access, camping coaches were located here both before and after the Second World War, the location being a popular one. At one time, three of these coaches were based here, located in the siding off the Ruabon line down platform.

58. Morfa Mawddach, 15th July 1964. Entering with the 8.5a.m. Machynlleth to Pwllheli train is BR Standard class 4-6-0 No. 75002. This was the last summer season in which the coast line services were shared with Llangollen line trains from Ruabon. There were a total of twenty eight mandatory stops on the two hour and fifty two minute journey and the possibility of another four calls should there be any business at Abertafol, Llanaber, Black Rock and Abererch halts. Connections were made at Dovey Junction – with through carriages for Aberystwyth – and Barmouth, for passengers intended for Dolgellau by means of the 9.45a.m. calling at Morfa Mawddach, Arthog and Penmaenpool. *Photo: C. L. Caddy*

60. Barmouth Junction, 3rd September 1959. GWR Mogul 2-6-0 No. 6335 arrives to a large number of passengers with a Pwllheli to Dovey Junction train during the last week of the summer timetable. Worthy of note are the comprehensive main buildings compared with the spartan facilities afforded on the Down platform. The Western Region enamelled nameboard is most informative but destined to be banished forever with the imminent renaming to Morfa Mawddach.
Photo: Michael Hale

59. Morfa Mawddach, 23rd May 1964. This time a southbound stopping train leaves the station with No. 75012 in charge. Forming the 3.25p.m. Barmouth to Machynlleth service, this working stopped at all stations except Abertafol which was conditional. Down trains were passed on the way at Llwyngwril and Aberdovey. The engine and stock had left Machynlleth in the morning as the 11'o' clock train to Barmouth (arrive 12.9p.m.) with conditional stops at Gogarth and Abertafol.
Photo: C. L. Caddy

61. Morfa Mawddach, 9th May 1961. The Western Region District Engineer from Oswestry would make a tour of his territory from time to time. Here, his one coach special (Saloon No. W80972) pauses at the coast line Down platform hauled by Ivatt class 2 2-6-0 No. 46519. Greetings are exchanged and points of interest raised. Notice the special headboard hanging from the smokebox handrail – reminiscent of the 'Radio Land Cruise' headboards of 1953. On the main platform, the station-master surveys the event. Within a couple of years, the whole line would be transferred from the Western Region to the London Midland Region and would come under the control of Crewe. The Inspection Saloon was subsequently preserved, and was available for private charter on the Cardigan Bay Express trains in 1987.

Photo: J. R. Besley

62. Morfa Mawddach, 23rd May 1964. A quite moment at the junction as all is prepared for the summer season. This view south shows the lines to Dolgellau and Ruabon bearing off to the left with the coast lines to the right.

Photo: C. L. Caddy

63. Morfa Mawddach, 29th September 1961. Croes Newydd based '43XX' class 2-6-0 No. 7339 draws into Morfa Mawddach with the 10.20a.m. from Barmouth to Ruabon. The fireman is ready to pick up the single line token from the signalman for the section to Penmaenpool, having surrendered the token for the Barmouth to Barmouth Junction section to the lineside apparatus. During weekdays this service carried through carriages from Barmouth to Birkenhead. Essentially however, it provided a semi-fast service to Ruabon, cutting out many of the wayside halts on the route. There were conditional arrangements for Llangower where the driver would stop the train on receipt of the necessary hand signal from passengers desiring to join. Llandderfel and Llandrillo passengers for Corwen, Llangollen, Ruabon and beyond, were to give advance notice of their intent to travel to the respective station staff. A connection was made at Bala Junction with the 11.25 ex Bala train, the 54½ mile journey taking two hours and twenty one minutes.

Photo: R. F. Roberts

64. (Above) Barmouth Junction, c.1930's. This official Great Western Railway photograph was taken to illustrate the Camping Coach, No. 9982 then in use, and located in the bay platform off the Llangollen line Down side. The vehicle is a six wheel clerestory coach. Access was by the portable ramp, and it was protected from collision by the sleeper just visible beyond the steps across the tracks. This short version camping coach was subsequently replaced by the longer bogie vehicles shown in other photographs.

Photo: GWR No B11099

65. (Centre) Barmouth Junction, 3rd September 1959. This view taken from the Dolgellau line down platform shows '74XX' class 0-6-0 pannier tank No. 7405 at rest with the Barmouth to Dolgellau train of one coach. These workings were very popular with the local people and greatly missed when the line closed from the 18th January 1965. Notice the difference in platform levels caused by reconstruction of a section based purely on the needs of local traffic.

Photo: Michael Hale

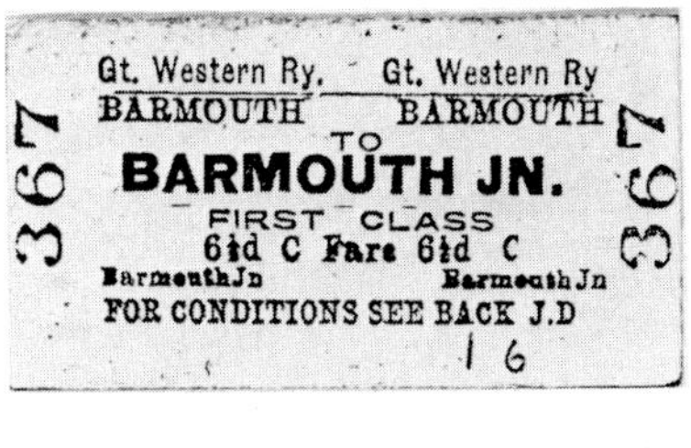

66. (Left) Barmouth Junction, c.1956. Pulling away from the platform into the Penmaenpool section, a member of the 'Manor' 78XX class No. 7817, *Garsington Manor* heads the 'North Wales Radio Land Cruise' on the next stage of its journey. Leaving Pwllheli at 10.10a.m., the train completed a circular tour of North Wales via Portmadoc, Dolgelley, Corwen and Rhyl, where a break was allowed of approx 2½ hours. Participants were transported back to their Cambrian line destinations via Bangor and Caernarvon, reaching Afonwen at 5.10p.m.

Photo: L. Hobdey

67. Morfa Mawddach, c.1963. Approaching the junction with a five coach stopping train to Ruabon, BR Standard class 4 No. 75023 blows off steam as she leans to the curve of the Dolgellau branch. The stock appears to be the result of an impromptu train marshalling exercise with a distinctly motley collection. The leading vehicle is a GWR bow end brake second followed by LMS 'porthole' corridor seconds, BR standard Mk1 composite, GWR full brake, and LMS full brake bringing up the rear. The train is possibly the evening 'Up mail' which left Pwllheli at 5.30p.m. and formed the 7.15p.m. Barmouth to Ruabon service with through carriages to Chester.
Photo: C. L. Caddy

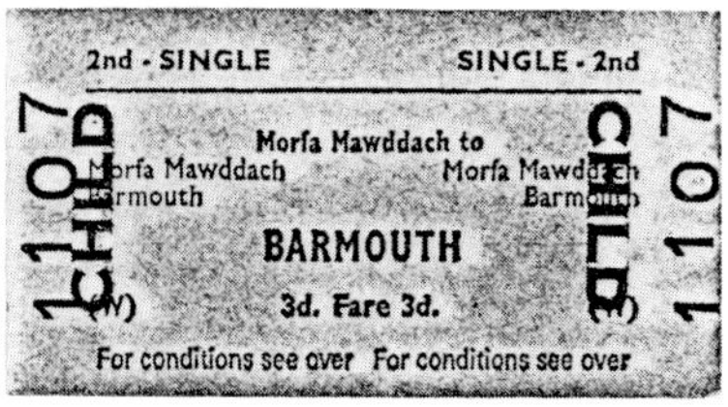

68. Morfa Mawddach, 7th June 1963. A view which captures the imagination of the exposed nature that was Barmouth Junction. The arrival of a train would see a flurry of activity as tokens for single line sections were exchanged. About to take the Dolgellau line with a Chester train is No. 75006. The platform has a largish number of passengers waiting whilst the signalman stands in readiness to hand over the token for Penmaenpool. The line to Fairbourne and subsequently Dovey Junction fades away to the right whilst the path running parallel with the train is known to many walkers and holiday makers alike, being the southern approach to Barmouth Viaduct.
Photo: B. J. Ashworth

69. Barmouth Viaduct, 3rd October 1964. A serene autumn day sees Ivatt 2-6-0 locomotive No. 46521 crossing the timber section of the viaduct with a Barmouth to Dolgellau train in the last months of the working. *Photo: M. S. Welch*

Barmouth Viaduct

Almost legendary in its own lifetime this famous structure carries but a single line across the Mawddach estuary. As far as official records are concerned, the bridge is really three in one with the 113 span timber trestle viaduct at 99 miles 41 ½ chains – from Whitchurch – being the first to be crossed on a northerly journey. Although opened for traffic in 1867, this section was reconstructed between 1906 and 1909. The second 'bridge', at 99 miles 74 ¼ chains, consists of three openings and includes the swing portion, completed in 1900 to replace the earlier 'overdraw' structure. The main features of this section are the hogback lattice girders with overhead bracing and the steel support cylinders. Finally the last portion of the viaduct is made up of two openings at the Barmouth end, at 99 miles 79 ½ chains, and consists of steel plate, cross and through girders, a reconstruction of 1900. Originally it was proposed to construct the viaduct using wrought iron columns but a number of factors, chiefly financial, resulted in the new famous timber structure. Baltic lumber was used, ships bringing the materials direct to site and unloading on the works. Of interest is a paper prepared in 1870-71 for the Institution of Civil Engineers by Mr. Henry Coneybeare, entitled 'Description of Viaducts across the Estuaries on the Line of the Cambrian Railway". The paper goes into great detail about the construction of two viaducts – across the Dovey and Mawddach estuaries – and the problems encountered during the period of building. In hindsight, Mr. Coneybeare's comment that Barmouth harbour was 'free from worms' appears to be a prophetic understatement following the events of the early 1980's when the viaduct was visited by a species of marine worm which almost spelt the death knell for the line. Fortunately, finances were found to save this magnificent example of our railway heritage.

70. (Right) Barmouth Viaduct, 31st May 1963. The slender outline of the viaduct can be appreciated in this view south across the Mawddach estuary from Panorama Walk, a vantage point above Barmouth Bay which was included in a booklet 'Rambles around the Cambrian Coast' by Hugh E. Page, and published by the Great Western Railway in 1936. *Photo: P. E. Baughan.*

71. (Above) Barmouth, c.1925. A south-bound passenger train crosses the timber trestle section of the viaduct reconstructed between 1906 and 1909. To the left of the picture is the original Borthwen or Orieltan viaduct, partially filled in 1928 and reduced to three spans. The low tide does not reflect the depth of the deep water channel below the swing bridge section which was – and is – upwards of fifty feet.
Photo: Miss Catherine Williams Collection

72. Barmouth, c.1963. Once the older locomotives had been withdrawn, the British Railways standard classes replaced them on the Cambrian Coast. The class 2 tender engines were considered to be under powered for some operations and the class 3 2-6-2T's took over a number of the workings. Here, 82021 comes off Barmouth bridge and crosses over the life boat slipway with the Dolgellau to Barmouth two coach shuttle. The locomotive's 6E shedplate indicates that it was nominally allocated to Wrexham (Croes Newydd) shed, so in all probability it was filling in working on one of the shed's cyclic rosters.
Photo: D. Cross

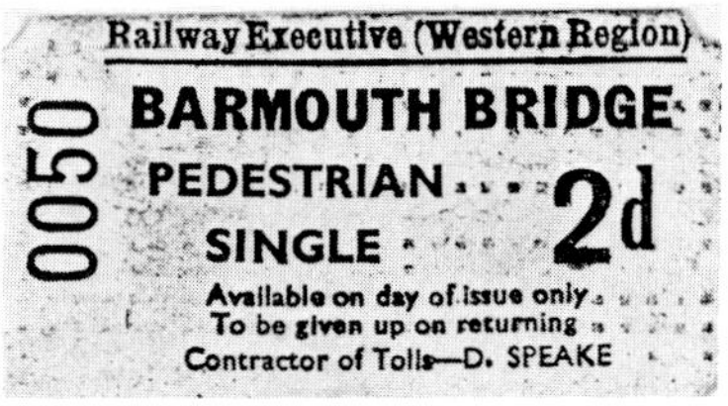

73. Barmouth, July 1964. Having left the viaduct with its 10 miles per hour speed restriction, BR Standard class 4-6-0 No. 75002 summons up the extra power needed to negotiate the slight rise in gradient into Barmouth. Compare this view with that at the top of the page to illustrate the filling in of Borthwen viaduct. Pedestrians descending the footpath to the left are more than likely seeking to take the advantage of the footpath crossing the viaduct, for which there was a toll charge.
Photo: T. Lewis

74. Barmouth, 3rd October 1964. On the last stage of its journey, Ivatt 2-6-0 No. 46509 approaches Old Chapel Viaduct with the 1.25p.m. Chester to Barmouth train. The train left Chester worked by Croes Newydd men, who took the train as far as Dolgellau before being relieved by a crew from Penmaenpool. Barmouth was due to be reached at 4.32p.m. The viaduct, of six spannings, was a pre-stressed concrete beam structure built in the 1950's to replace a life expired timber bridge.
Photo: M. S. Welch

BARMOUTH

Barmouth station, 22 miles 32 chains from Dovey Junction and the most important intermediate station on the fifty four miles from Dovey Junction to Pwllheli, passenger trains taking just over the hour for the journey. Most trains from the Dolgellau and Ruabon line terminated here, as did some from Pwllheli working south and Machynlleth working north.

On leaving the bridge proper, the line swings through a seventy degree turn, initially across a short masonry faced embankment and the lifeboat slipway. A small timber underbridge, known locally as Borthwen, is next crossed as the railway enters a short shallow tunnel over which runs the coast road into the town. To the left the passenger now sees Barmouth harbour as the train crosses Old Chapel Viaduct, a six span concrete structure, and a low road bridge before bearing right to enter a double line section under the control of Barmouth South signal box. Although the home signals are passed, the single line staff is not surrendered until the box is reached.

The station layout was broken by a level crossing and comprised of the main Up and Down platforms, passengers crossing the line by a footbridge just north of the level crossing. There was a north facing bay platform line off the Down side. South of the crossing on the inland side was an excursion platform, regularly used for the Dolgellau shuttle and for the one Sunday train from Machynlleth, a practice which avoided the necessity to open Barmouth North signalbox on Sunday. The section from Barmouth South to Barmouth North was worked as double line and signalled accordingly by the Absolute Block system. The level crossing was worked from Barmouth South box and controlled by interlocking signals. The opening of the crossing gates to trains caused considerable congestion to road traffic in the summer months and the problem still exists to this day.

All storage lines and sidings were located north of the station and there were facilities for train crews to water, service, and turn locomotives. At one time a 45ft diameter turntable was located on the down side, but was taken out of use some time before the removal of surplus sidings during the early 1970's. The freight sidings were located on the Up side with a goods shed behind the Up platform. Barmouth North signal cabin was situated beyond the Down platform ramp.

The main station buildings, constructed of dressed stone, were located on the Up (No. 3) platform and had a large canopy extending over the platform affording some protection to passengers. The Down (No. 2) platform shelter was a much simpler affair but also had a canopy which extended the full width. The Down bay platform, (No. 1) had no shelter. Barmouth could be a very bleak location even in the summer months. At the extreme north end of the station limits, a footbridge crossed the tracks, providing an alternative means of getting to the promenade.

The line was restricted to 'Yellow' group locomotives until 1938 when bridges and other structures were upgraded, subsequently enabling the passage of 'Blue' class locomotives. There were some limitations even so, and the Sectional Appendix to Section 16 Working Timetables listed the restrictions. Cambrian engines gradually gave way to the older GWR designs which tended to be hand-downs from other lines who had received newer locomotives. Even after nationalization the habit persisted and it was only with the advent of the BR Standard class 2 78XXX tender design 2-6-0's that Machynlleth, Portmadoc and Pwllheli finally received brand new locomotives. Whilst the 32XX class 4-4-0 tender engines were classified as new designs, they were in reality a rebuilt hybrid incorporating parts from much older locomotives.

It is somewhat surprising that no locomotive depot was established at Barmouth in view of the distance between sheds. In all probability the 44 miles between Machynlleth and Portmadoc was one of the longest sections on single lines in England and Wales between depots, and although there was the small shed at Penmaenpool, provision was restricted to providing locomotives and men for the Dolgellau to Barmouth shuttle, and, after amalgamation, between Barmouth and points on the Ruabon line.

75. (Above) Barmouth, c.1959. British Railways standard class 2 tender engine No. 78007 pulls out of Barmouth with a train for Dovey Junction, and heads towards bridge with a class A train of assorted stock. The leading coach is a non corridor third, whilst the next two are full brakes. The locomotive sports the later design of crest on the tender. Probably the working is the 3.45p.m. from Pwllheli, which carried Mails to Dovey Junction where the stock was transferred to the 6.00p.m. from Aberystwyth to Shrewsbury. No. 78007 was nominally based at Machynlleth but spent a lot of its time at Portmadoc or Pwllheli. *Photo: D. Cross*

76. (Below) Barmouth, July 1953. During 1953, the stretch of line between Barmouth tunnel and Barmouth South signal box was taken over by the District Engineers who were engaged in a programme of rebuilding Old Chapel Viaduct, the low six span bridge that skirted the waterfront by the harbour at Barmouth. Once the morning train from Machynlleth was safely stored in the excursion platform, the engineers took possession of the line from about 10.15a.m. until 4.15p.m. and performed their programme of work, which lasted for several months. Here, 4-6-0 Manor class No. 7826 *Longworth Manor* of Croes Newydd shed (84J) stands silently whilst a rail mounted crane performs its duties in the background. In the distance can be seen the mouth of Barmouth tunnel with the mountain tops shrouded in mist in the distance. *Photo: W. G. Rear*

77. Barmouth, c.1951. On entering the town, a double line section was encountered under the control of Barmouth South signal box. A Down train is being held between the home signal and level crossing as both platforms in the station appear to be occupied. The Sunday workings worked into the up bay platform seen on the right. The view of Barmouth promenade speaks for itself as it awaited development to cater for the motor car age. *Photo: J. H. Moss*

78. Barmouth, c.1956. 4-4-0 'Dukedog' No. 9017, now preserved and running on the Bluebell Railway seen here on its former stamping grounds, runs into Barmouth with a four coach working from Dovey Junction. Just visible is a 43XX class 2-6-0 standing in the excursion platform. Between the Down signal and the signal box can be seen a Crosville double decker bus, probably one of the Leyland TD7 type which resided at Barmouth for many years, working the competing route to Dolgellau. *Photo: J. W. T. House*

79. (Below) Barmouth, 7th August 1959. GWR Collett 0-6-0 No. 3201 of 84J shed draws out of Barmouth Station with a class B train which includes SR and LMS stock. These locomotives provided the power for much of the traffic along the coast, which they shared with the older 4-4-0 'Dukedogs' and the smaller 2-6-2T engines of the '4575' class. *Photo: R. W. Hinton*

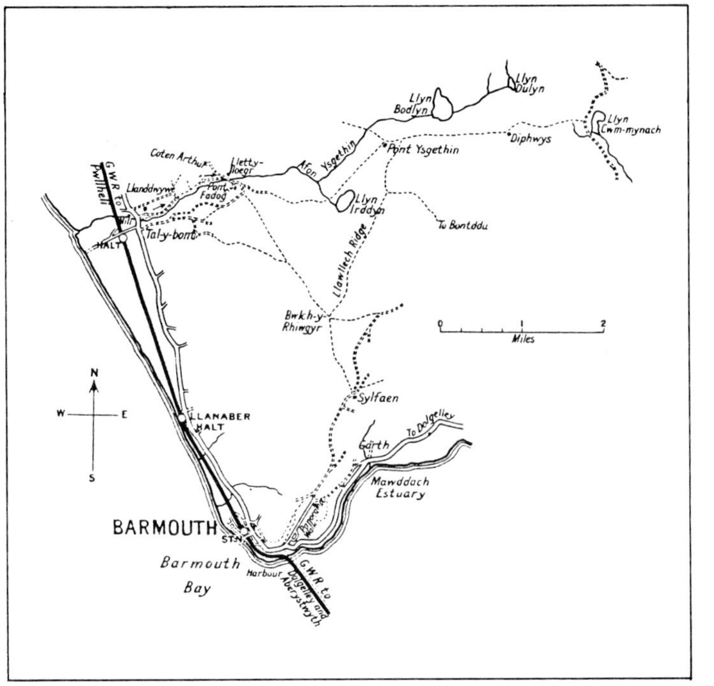

Above: Reproduction of map detailing rambles from Barmouth which include Llanaber, Tal-y-Bont, The Ysgethin Glen, Diphwys, The Panorama Walk.

80. (Left-top) Barmouth, 3rd July 1969. Although the motive power changed over the years and steam had long disappeared from the Cambrian Coast, Barmouth South signal box remained unchanged. Note the tablet receiver arm, protected by the rope net, with the lamp on its small post to illuminate it at night. The box contained a 27 lever frame.
Photo: C. L. Caddy

81. (Centre) Barmouth, 1963. Barmouth station exterior, about 1963, showing the main entrance hall, and the parcels office doorway. In the distance can be seen the shelter for passengers using the excursion platform, whilst a representative collection of vehicles stand in the forecourt. A member of the station staff chats to a potential customer in the yard.
Photo: G. H. Platt

82. (Left) Barmouth, 13th July 1964. Another view of the station exterior, which leads on to the up platform. An abundance of notice-boards of varying sizes are put to good use and the usual contempt for 'No Parking' notices was prevalent even then – unless of course the vehicles belonged to station staff. *Photo: C. L. Caddy*

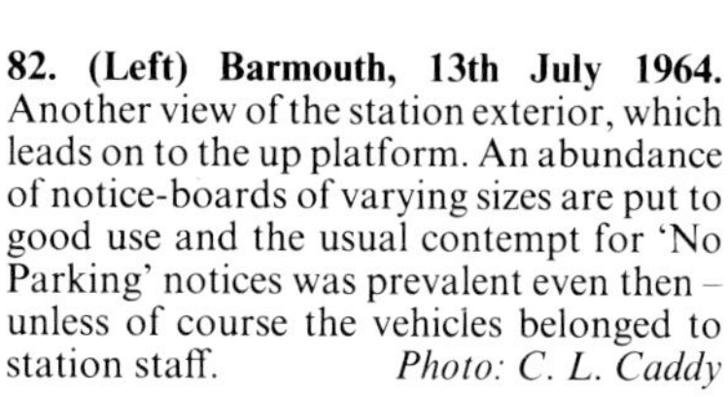
1085 GT. WESTERN RLY GT. WESTERN RLY 1085
HALF DAY EXCURSION
RETURN TICKET
Barmouth Barmouth
TO 8.00
DOLGELLEY
AND BACK
THIRD CLASS
Dolgelley Dolgelley
FOR CONDITIONS SEE BACK W.D

83. Barmouth, c. 1900. Cambrian Railways station staff at Barmouth gather for their group photograph. Unfortunately, no date is stated and the identities of the gentlemen remain unknown. However, certain members have details of their employment on either cap or coat lapels. On the back row, extreme left, is the Town Porter. Front row, extreme left, is the Foreman. Second from right, sitting down, is one of the two Ticket Examiners, the other being on the back row in front of the open door. On the extreme right, standing and with a button-hole, the employee has the legend 'Cambrian Railways' on his lapels. Numerous others with the oval lapel badges indicating their status are within the group. Presumably, the gentleman in the centre of the front row is the Station Master.
Photo: Miss Catherine Williams Collection.

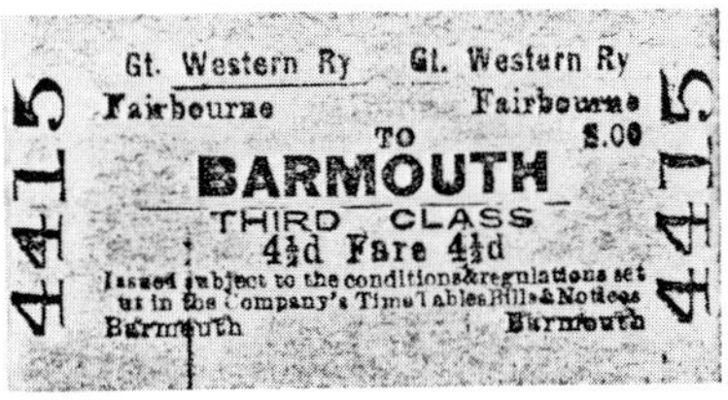

84. Barmouth, c.1963. A view looking south along the Down platform. Two coaches have been detached from a north bound train and the Dolgellau shuttle locomotive will collect and dispose of them in due course. Groups of passengers congregate beneath the canopy whilst awaiting their connection.
Photo: Lens of Sutton.

84a. Barmouth, c.1968. Looking north from the up platform towards Harlech, a six-car diesel multiple unit recedes into the distance after passing Barmouth North signal box, a thirty-six lever frame which had closed by July 1975 and replaced by a ground frame. *Photo: C. L. Caddy*

85. Barmouth, 3rd October 1964. British Railways Standard class 4 2-6-4T No. 80104 pulls away from Barmouth Station with a Pwllheli train. Of all the 'Standard' locomotives that worked on the line, these were regarded with affection by many of the crews. They were masters of the job and would pull almost anything that was attached to them, either passenger or freight stock. The cabs were comfortable and crossing Barmouth bridge, either direction, gave a degree of shelter from the elements that had been conspicuous by its absence in the past. The locomotive diagrams for this class of locomotive meant that these locomotives worked three single trips each day. Taken from the footbridge north of the station this view shows how the town has literally been built into the hillside.
Photo: M. S. Welch

86. Barmouth, 3rd October 1964 (Saturday). The same train as above, now having passed the photographers, sets out for open countryside on its journey to Pwllheli. During the week this working provided a service to the schools at Barmouth and Harlech and stopped at all stations and halts en route with the exception of Black Rock, which was conditional. The goods yard has by now closed for public use.
Photo: M. S. Welch

LLANABER HALT

Located 1 mile 34 chains from Barmouth, passenger trains generally took 3 minutes for the journey. The course of the line was generally flat, in fact for about a mile the gradient fell at 1 in 1754, but a short climb of 1 in 100 preceded the entry to the platform, which was on the level.

The halt was located on the Up side, the platform length being 202 feet, and comprised of a sleeper edged platform, with a small wooden shelter, together with a standard station name board and a platform oil lamp. Access was by footpath from an occupation road. Guards collected/issued tickets, although the station came under the control of the Stationmaster at Barmouth. The halt is still in use today (1988) serving nearby caravan sites.

87. (Right-upper) Llanaber Halt, 13th July 1964. Llanaber Halt, looking towards Harlech, showing the wooden edged platform, name-board, shelter and the occupation crossing beyond, which led to the beach. The station was sheltered by the high ground behind the platform, but nevertheless was a somewhat bleak location when the weather was bad – even in summer. At the time of the photograph, seven Up trains were booked to call, one conditional. In the Down direction, seven were also booked to call, three being conditional.

Photo: C. L. Caddy

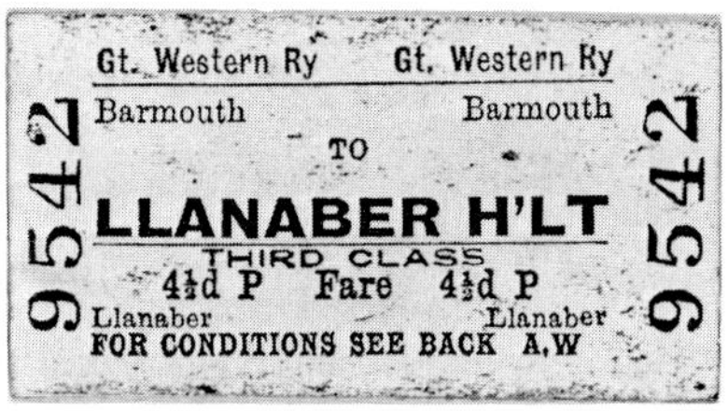

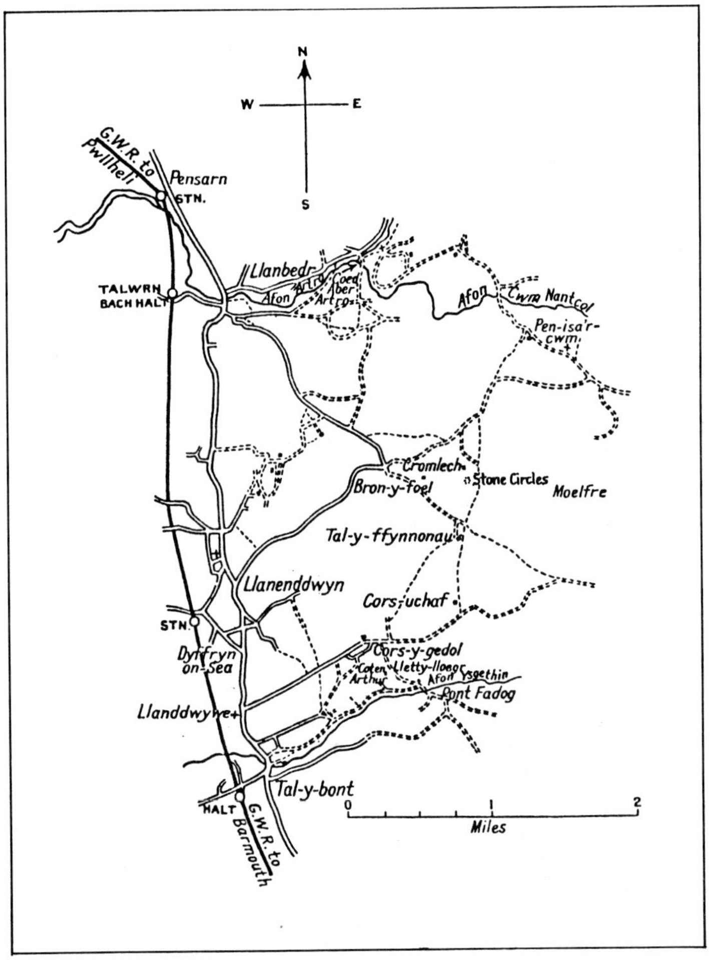

TALYBONT HALT

Talybont Halt is 3 miles 58 chains north of Barmouth, and local trains took seven minutes for the journey. The track was subject to several changes of gradient between Llanaber and Talybont, rising and falling over very short distances. The climb up to Talybont Halt was at 1 in 80, with the platform face, 205 feet in length on the Up side. The usual construction methods had been employed, old sleepers edging the platform with infill topped with chippings. The platform rear had a post and rail fence to protect passengers and the shelter was of corrugated sheet construction with windows on the ends.

A standard name board and lamp completed the furniture. Beyond the station, a minor road crossed the line by means of a plate girder construction bridge. Access to the station was by a footpath off the road and passengers walked under the bridge to get to the platform. There was some protection from the sea in the form of a small rise which had a few trees affording some shelter. The station is still open today (1988), as ever an unstaffed halt, with the guard performing the role of fare collector. Talybont also came under the control of Barmouth Stationmaster although it was nearer to Dyffryn Ardudwy.

88. (Right) Talybont Halt, 13th July 1964. Taken from the track level and looking towards Harlech, the platform looks to have some shelter from the trees on either side of the line. The waiting hut seems devoid of comfort, although no doubt there would be some protection from the worst of the elements. There is a bench seat outside the hut, which was probably as comfortable as the fittings inside. Beyond the hut can be seen the holder for the oil lamp, whilst the 'way out' sign is visible at the end of the ramp, which is under the road overbridge. In the distance can be seen a platelayer's hut, and the line falling away as the gradient changes.

Photo: C. L. Caddy

DYFFRYN ARDUDWY

Originally Dyffryn – renamed Dyffryn-on-Sea after the Grouping (to avoid confusion with Dyffryn in the Rhondda Valley). Renamed Dyffryn Ardudwy, this was the first passing point north of Barmouth, 4 miles 71 chains south. Passenger trains were allowed ten and a half minutes for the journey. The Up loop line was 684 feet in length whilst the Down line was 705 feet. The station was on a rising gradient of 1 in 100 from Barmouth and was renamed from Dyffryn-on-Sea on 1st June 1948. There was a small goods yard with two sidings at the south end on the Up side, access being controlled from the 13 lever signal box which was constructed of timber with corrugated sheet cladding on a brick base and located on the Up platform. Housed here were token instruments for the Up line to Barmouth and the Down line to Llanbedr. Local instructions for station working were limited to ensuring that freight trains stopping to work at the station must stand clear of the level crossing, which was located at the south end of the station, responsibility for this being carried out resting with the guard of the train concerned. In view of the gradient it was necessary to ensure that sufficient wagon brakes were pinned down to hold stock safely. Goods facilities were withdrawn on 2nd December 1963. The Down line through the station was removed when the passing loop was taken out. Today, the station survives as an unstaffed halt.

The main buildings were on the Up platform, the structure containing the Stationmaster's house plus the usual offices. There was a small canopy giving limited protection but passenger traffic was generally sparse and the Cambrian did not throw money about in isolated locations. The Down platform had a wooden shelter at the northern end of the loop, and passengers crossed the line by a walkway across the tracks at the Harlech end. In the goods yard was a store, a cattle pen and loading dock, a small shed, and coal tenancies, with a weigh bridge and office by the road access to the yard.

89. (Right-upper) Dyffryn Ardudwy, c.1958. The up platform at Dyffryn Ardudwy looking towards Barmouth. There appears to be an excess of baggage handling equipment visible and the one passenger seems content to stand rather than use the bench seats provided. The change of gradient is apparent as the track disappears towards Barmouth. *Photo: Lens of Sutton*

90. (Centre) Dyffryn Ardudwy, 13th July 1964. The down platform, looking towards Harlech. The shrubs provided an attractive cover in the season and the station was always kept neat and tidy by staff who took a pride in their work-place. The small signal cabin appears to be constructed of corrugated sheet in this view, although other views in this work indicate that the other side was constructed of timber. Possibly the box was repaired at some time between the taking of the photographs. The platform height was rather low, and steps are visible on both platforms. It was the practice to have several sets of steps to assist passengers alighting and boarding at stations with low platform heights. Notice how the gradient falls away as the track heads north. *Photo: C. L. Caddy*

91. (Right-lower) Dyffryn Ardudwy, circa 1958. British Railways standard class 2 2-6-0 No. 78002 heading a three coach train coasts into Dyffryn Ardudwy, one afternoon in 1958. The signalman waits to collect the train staff for the section from Barmouth. It appears that he has no staff to offer for the next stage, so possibly another train is going to cross here. Apart from railway employees there is a distinct shortage of passengers. *Photo: C. L. Caddy collection*

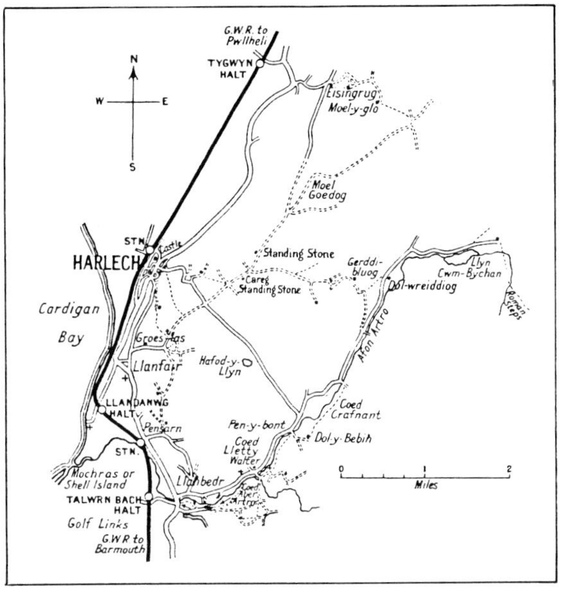

TALWRN BACH HALT

Seven miles 8 chains from Barmouth was Talwrn Bach Halt – renamed Llanbedr in 1978 – in the Dyffryn to Llanbedr & Pensarn (the original) section and on a level stretch of track. The 101 foot long platform, which came under the control of the Stationmaster at Llanbedr & Pensarn, was located on the Down side, the original wooden faced platform being replaced by a modern concrete supported slab structure. A standard GWR design corrugated sheet shed served as shelter, with a name-board and two oil lamp supports to complete the furniture. At the Barmouth end of the station a public road crossed the track, with the gates normally set across the track and a crossing keeper in attendance. The gates were protected by signals. At the Harlech end of the platform is a platelayers cabin.

Down trains took about fifteen minutes to reach the halt from Barmouth, the guard issuing or collecting tickets as required.

92. (Right-upper) Talwrn Bach Halt, c. 1960. Opened by the GWR on 9th July 1923, Talwrn Bach Halt was situated on a long straight section of track – one of the few on the coast line. In this view, looking towards Barmouth, the crossing keeper's cottage and the level crossing gates in their usual position across the tracks are in evidence. *Photo: Lens of Sutton*

93. (Centre) Talwrn Bach Halt, 13th July 1964. The short length of the platform is emphasised in this view, as is the modern pre-cast concrete supports for the platform although this is in contrast with the traditional GWR designed shelter. Nearby is the Royal Aircraft Establishment at Llanbedr, where pilotless aircraft are flown. *Photo: C. L. Caddy*

94. (Right) Llanbedr & Pensarn, 6th October 1972. On the southern approach to Llanbedr & Pensarn Station the railway crosses Afon Atro on a low timber trestle bridge. The 4.18 p.m. DMU from Machynlleth to Pwllheli crosses this structure forming a picturesque background to the small estuary moorings for leisure craft. *Photo: C. L. Caddy*

LLANBEDR & PENSARN

Llanbedr & Pensarn station, now renamed *Pensarn,* was the next token exchange station after Dyffryn Ardudwy, 7 miles 67 chains from Barmouth. However there was only one platform face, on the Up side. The Up loop line was 684 feet in length, whilst the Down line was 744 feet. Freight facilities were limited to a loop off the Down line and contained a short siding with a cattle dock and pens at the Barmouth end. A hand operated level crossing at the southern end of the platform completed the picture.

The buildings on the platform comprised of a signal cabin with 34 lever frame, Stationmaster's house with adjoining offices for station duties and a store and lamp room. Between the Down loop line and the goods line a small warehouse once stood, but which had been removed before nationalization.

Whilst there were no restrictions on Up and Down passenger trains crossing at Llanbedr, it was arranged that this never happened. Freight trains however would pull into the Down line and so both lines through the loop were signalled for either direction. Local instructions stated that trains using the Down loop must stop to change staffs.

Passenger trains were allowed 18 minutes from Barmouth. After the demise of steam, the loop and goods lines were removed and the station reduced to an unstaffed halt on the section between Harlech and Barmouth.

95. (Left-upper) Llanbedr & Pensarn, c. 1960. Taken from the platform looking south towards Barmouth, the hand operated level crossing with protecting signals is beyond the platform ramp. In the distance can be seen the small trestle bridge over the creek. Notice the lower quadrant signals, the standard GWR design sheds, the tablet catcher beyond the level crossing and the large nameboard. *Photo: Lens of Sutton*

96. (Centre) Llanbedr & Pensarn, 13th July 1964. Taken from the level crossing at Llanbedr, looking towards Harlech and Pwllheli, this view shows the station buildings, both sheds and beyond the signalbox, which dates from the 1930's, when it was erected to replace the former Cambrian Railways 12 lever box which was located by the level crossing. The station was signalled for trains to work in either direction – through the loop or the platform line. At the foot of the ramp can be seen the token delivery holder. *Photo: C. L. Caddy*

97. (Left-lower) Llanbedr & Pensarn, c. 1950. Four wagons stand on the goods line at Llanbedr which was little used. One van is somewhat unusual in that it was used for carrying motor cars. It is unlikely that it was performing this function here. The wagons would have to be pinned down to avoid running away, for the station was on a falling gradient towards Harlech of 1 in 284. *Photo: J. H. Moss*

NORTH WALES RADIO LAND CRUISE

SPECIALLY EQUIPPED FOR ACTUAL RADIO RECEPTION AND DESCRIPTIVE COMMENTARY ON FEATURES OF INTEREST EN ROUTE.

FOR MAP OF ROUTE PLEASE SEE OVERLEAF.

TUESDAYS and THURSDAYS
5th JULY to 21st JULY, 1960 (inclusive)
ALSO
MONDAYS to FRIDAYS
25th JULY to 2nd SEPTEMBER, 1960 (inclusive)

FARE	Time of departure	FROM	Arrival on Return
s. d.	a.m.		p.m.
13/- FROM EACH STATION	10 10	PWLLHELI	5 C28
	10 18	PENYCHAIN	5 C13
	10 33	CRICCIETH	5 14
	10 43	PORTMADOC	5 23
	10 50	PENRHYNDEUDRAETH	5 34
	10 55	TALSARNAU	5 42
	11 0	HARLECH	5 46
	11 6	LLANBEDR & PENSARN	5 58
	11 12	DYFFRYN ARDUDWY	6 0
	11 15	TALYBONT HALT	6 3
	11 25	BARMOUTH	6 10
	11 45	PENMAENPOOL	7 *36
	11 50	DOLGELLEY	7 *41
	11‡25	BALA...	8†*42
	p.m. 12 50	CORWEN	9 * 2

RHYL arrive 1-58 p.m. RHYL depart 3-30 p.m.

*—Change at Barmouth C—Change at Afon Wen
‡—Change at Bala Junction and Corwen. †—Change at Bala Junction.

Children under Three years of age, Free; Three and under Fourteen years of age, Half-fare.

LIGHT MEALS AND REFRESHMENTS, MAY BE OBTAINED FROM THE CAFETERIA CAR ON THE TRAIN.

HOLIDAY RUNABOUT TICKETS ARE NOT AVAILABLE BY THIS TRAIN

For conditions of issue of these tickets, also luggagr allowances, see the British Transport Commission's Regulations and Conditions of Issue of Tickets, etc.

Further information will be supplied on application to the Stations, or to Mr. O. VELTOM, District Traffic Superintendent, Oswestry (Telephone Oswestry 2291, Extn. 211); or to Mr. R. L. CHARLESWORTH, Commercial Officer, Paddington Station, W.2.

BRITISH RAILWAYS

ROUTE OF THE NORTH WALES RADIO LAND CRUISE TRAIN

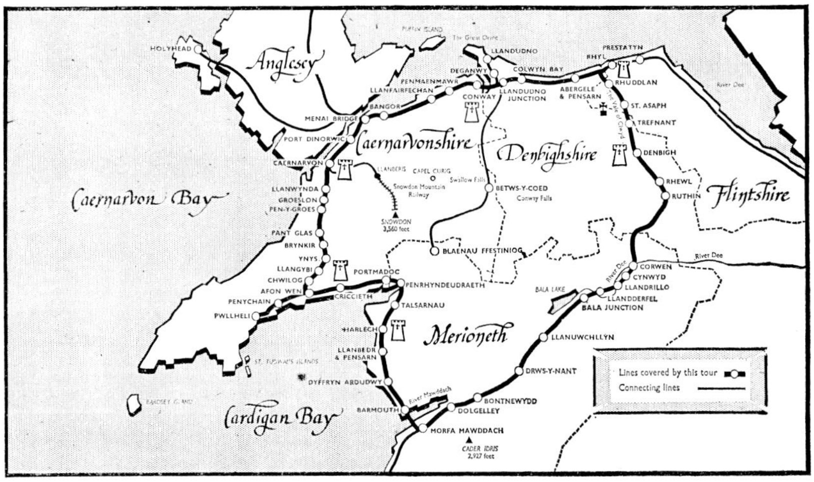

Published by British Railways (Western Region) No. 63. Printed in Great Britain by G. R. Griffith Ltd. Chester

Above and Left: Half size reproduction of the publicity handout produced for the 'North Wales Radio Land Cruise'. A special descriptive itinerary was given to the passengers on the train which for many years was a feature of holidays in the area.

LLANDANWG HALT

Fifty one chains after Llanbedr & Pensarn towards Pwllheli, and eight miles and 38 chains from Barmouth, we arrive at Llandanwg Halt. This unstaffed halt comprises of a single platform 75 feet long and is located on the Down side on the section between Llanbedr & Pensarn and Harlech, the halt coming under the control of the Stationmaster at Llanbedr. The platform edge is faced with wood, the rest of the platform having earth topped with a surface coat of chippings. Up trains approached the station blind, sighting of the platform being restricted by a road overbridge north of the platform. The station furniture was very basic, with one corrugated sheet shelter, open to the elements but with its back to the sea. An oil lamp and nameboard complete the inventory. Access to the platform was by an inclined footpath from the nearby road which continues to serve a small number of cottages and a popular beach.

At this point the track was level, but beyond the station travelling north, the gradient steepened for a short distance at 1 in 75.

98. (Centre) Llandanwg Halt, 27th June 1968. Taken from the road bridge at the north end of the halt, looking towards Barmouth, the track follows the contour of the shoreline on a falling gradient as it progresses south. *Photo: C. L. Caddy*

99. (Right) Llandanwg Halt. A view of Llandanwg Halt, looking north towards Harlech and Pwllheli. The small platform appears crowded, for in addition to the furniture mentioned in the text are two disused lamp posts and a 'trespass warning' sign. As if that wasn't enough, a speed restriction sign adorns the platform at the north end. Clearances at the overbridge look very tight. The halt was opened by the GWR on 18th November 1929. *Photo: C. L. Caddy*

100. (Above) Harlech, c. 1950. Livestock market day involved much activity for farmers and the railways, and the alternate Wednesday market was of great importance to the community. The 'Dukedog' from Portmadoc shed was occupied shunting the cattle trucks, for the small loading bay and pens limited the amount of animals that could be dealt with at any one time. Pre-war design cars are much in evidence, and apart from the Morris "E" in the left hand foreground, most of the other vehicles are of pre-war design. A Bedford 'Utility' cattle lorry stands in front of the pens in the right hand foreground. Notice the enamel sign for 'Astons' on the Down platform. Every station in North Wales carried at least one advert for this firm, now alas swept up in amalgamation and take-over deals. Of particular interest are the two coach bodies serving as offices and stores, shortly to be replaced by a modern structure, and the open landscape behind the station. The council houses have recently been constructed and the exposed nature of the ground towards the sea offers little protection from the elements. *Photo: J. H. Moss*

HARLECH

Harlech lies 10 miles 33 chains from Barmouth, the one point between Barmouth and Portmadoc where passenger trains can cross. Running time for class B passenger trains was 27 minutes from Barmouth. The station lay under the shadow of Harlech castle, perched high on a rocky outcrop. The line descended into the station on a gradient of 1 in 330 for about half a mile, then levelled off for about 200 yards before falling again for a further mile at 1 in 660 to beyond the station where it levelled out. The track extended in a straight line for about four miles.

Initially there was only a platform on the Down side, but the Up side platform was added later. The Up loop was 837 feet in length, whilst the Down loop was five feet longer. There was a refuge siding doubling as a shunting spur which could accommodate 40 wagons plus engine and brake van on the Up side.

The main buildings were located on the Down platform, these including the Stationmaster's house and offices. One or two corrugated sheet sheds were located on this platform. On the Up platform, a GWR design 'pagoda' shelter sufficed. A new footbridge was provided by the GWR for passengers. South of the platforms were two level crossings. A signal cabin with 27 levers on the Down side controlled the set of gates nearer the station and were normally set across the road and where the staff token exchange equipment was located. A secondary road crossed the line a hundred yards further on, known as Morfa level crossing, and was protected by signals and with a crossing keeper in residence to attend to the duties. These gates were normally set against the railway.

A goods loop was located on the Up side behind the Up platform and at one time, two coach bodies mounted on blocks were used as goods offices and stores, but these were removed and replaced with a single pre-cast concrete structure. Auctions were a regular event until the early 1950's, but this traffic diminished. There was also some coal traffic inward and general merchandise.

Excursion traffic used to run to Harlech from Barmouth on Sundays before the war in the peak season. It was a regular crossing point and for one or two seasons it was not unknown for footplate crews to change over here and work back to their respective depots. There is still some school traffic, and indeed the trains that catered for pupils was one of the deciding factors in retaining the line during the branch line purge of the Beeching era.

101. Harlech, 13th July 1964. From the footbridge looking south this view shows the attractive station buildings, the signal box by the station level crossing, the Morfa level crossing beyond. Access from the goods line behind the Up platform to the Up loop was just beyond the platform ramp. The immaculate appearance of the station indicates the pride that staff had for their work. *Photo: C. L. Caddy*

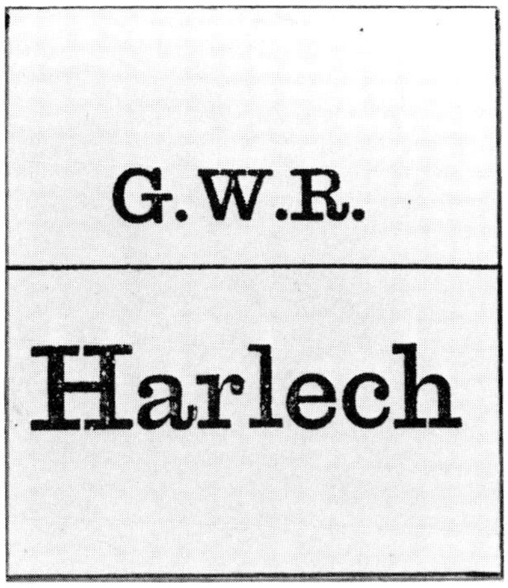

102. Harlech, 30th July 1963. British Railways standard class 3 2-6-2T No. 82032 arrives at the station with a Pwllheli to Dovey Junction passenger working. The signalman walks along the track towards the platform with the token in his hand for the section to Llanbedr & Pensarn. The open footbridge and pagoda type shelter show the GWR influence. On the Down side, a British Railways lorry stands by the offices loading up local deliveries, a service which regrettably has disappeared.
Photo: H. B. Priestley

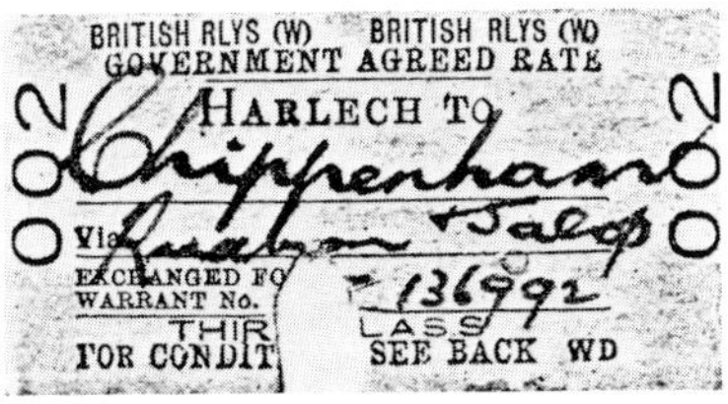
BRITISH RLYS (W) BRITISH RLYS (W)
GOVERNMENT AGREED RATE
HARLECH TO
Chippenham
Via Ruabon + Salop
EXCHANGED FO
WARRANT No. 136992
THIR LASS
FOR CONDIT SEE BACK WD

103. Harlech, 30th July 1963. Taken from the castle which dominates the entire area, a panoramic view of the station is provided emphasising the four mile straight stretch of track to the north as the line crosses Morfa Harlech. A five coach train bound for Pwllheli enters the single line heading northwards, having crossed a south bound train at the platform hauled by British Railways standard class 3 2-6-2T No. 82032, a familiar locomotive on the coast line before being transferred to Bangor shed and ultimate scrapping. Freight traffic had diminished and there is little sign of activity here. In another year, the goods yard would close completely.
Photo: H. B. Priestley

104. Harlech, c. 1911. The Cambrian Railways, ever since their formation, were always keen to advertise the Principality which they served and consequently small towns and villages achieved popularity with English tourists. Here 2-4-0 No. 53 arrives at Harlech from the south and disgorges its passengers. The locomotive was built by Sharp Stewart of Manchester in October 1865 (maker's number 1635) and was rebuilt in 1892. Although it lasted until the grouping and was allocated the GWR number 1322, it never carried its new designation, being withdrawn in August 1922. *Photo: A. S. Wilson Jones, courtesy M. S. Welch*

105. Harlech, c. 1911. A Pwllheli bound train draws to a halt at Harlech in the summer of 1911. The 4-4-0 locomotive was number 11 and was one of two built by the Cambrian Railways themselves at Oswestry in July 1904. It survived the amalgamation being renumbered 1068 by the GWR, but was withdrawn in January 1924. It will be noted that there is no Up platform, this facility being provided by the GWR when the station was remodelled. *Photo: A. S. Wilson Jones, courtesy M. S. Welch*

Great Western Railway.

THIRD CLASS

Holiday Season Ticket

This Ticket is issued subject to the Conditions and Regulations contained in the Company's Time Tables, Bills and Notices, is

NOT TRANSFERABLE

and must be given up on or before the date of expiry.

106. Harlech, c. 1911. The Cambrian Railways had the misfortune to suffer its share of minor accidents and mishaps from time to time, as well as having more than its fair share of major disasters. Here a Down goods was shunted into the Up loop to allow the last Down passenger to proceed. When the driver finally received the staff token for the next section, he was unaware that the signalman had failed to re-set the points. Furthermore, the driver was unaware or careless in his observance of the situation to realise that he was proceeding into the refuge siding and came into contact with the wagons there with some force. Neither of the crew were seriously injured, but doubtless they suffered some shock. The locomotive, No. 79, was built by the Vulcan Foundry in 1894 and survived into GWR days. *Photo: A. S. Wilson Jones, courtesy M. S. Welch*

TY GWYN HALT

Twelve miles 75 chains from Barmouth, and with class B passenger trains taking about 32 minutes to reach the halt, Ty Gwyn Halt is on the Down side of the track on the section between Harlech and Penrhyndeudraeth. A level crossing which was normally set across the tracks and protected by signals was controlled from a small three lever frame. A crossing keeper lived alongside the crossing and attended to the opening and closing of gates.

Ty Gwyn was opened by the GWR on 11th July 1927 with the very basic layout found at other halts on this line. The timber edged platform is 71 feet long and comes under the control of the Stationmaster at Penrhyndeudraeth. A corrugated sheet shelter facing inland provides basic shelter, with lamp standard and nameboard to complete the furniture inventory. Access is by a gate off the road. The track is still on the level on the long straight section between Harlech and Talsarnau.

107. Ty Gwyn, 25th May 1964. This uncomplicated facility appears to be situated in the middle of nowhere but springs to life during the summer months when holiday makers fill the nearby caravan sites. This view looking north shows the basic facilities described above. *Photo: C. L. Caddy*

TALSARNAU

Talsarnau is 13 miles 63 chains from Barmouth and an intermediate station on the section between Harlech and Penrhyndeudraeth. Class B trains were allowed 34 minutes from Barmouth. Although the station was not a staff token exchange point, platform staff were on duty throughout the day until it was reduced in status to an unstaffed halt during the recession cuts imposed on the line.

The station had one siding, connected with the running line, and faced Harlech. The points were worked by a ground locking frame with eleven levers and located in a small signal box at the Harlech end of the platform. Local instructions stated that the siding could be worked by both Up and Down freight trains. Shunting engines could be worked out from Penrhyndeudraeth and if necessary could propel wagons back there provided there was a brake van leading on the return journey with the guard or shunter keeping watch. Freight traffic was minimal, and after the Second World War, the siding was used to house a camping coach in the holiday season. Access to the siding was interlocked with signals. At the Portmadoc end of the platform was a level crossing controlled by a gatekeeper in residence, with the gates normally set across the road.

The platform was located on the Up side of the track, the station building being a substantial structure, of local stone, and incorporated the Stationmaster's accommodation.

108. Talsarnau, c. 1955. The Moelwyn range of mountains lie to the north of Talsarnau and are seen here in the distance. The Camping Coach is an ex GWR 'Toplight' vehicle. At the time it could be hired for about 12 shillings per week and was extremely popular with families, accommodating up to eight persons. If there was a snag however, it was that families who rented the coach had to travel by rail to and from their home station. At the far end of the platform can be seen the level crossing with its protecting signals. The siding to the rear of the camping coach catered for the small amount of freight handled there. Withdrawal of these facilities took place on the 4th May 1964. *Photo: J. H. Moss*

109. Talsarnau, 13th July 1964. This view, looking towards Barmouth, shows the simple design incorporating a small canopy to protect passengers from the worst of the weather although the station in fact faced the sea. Beyond can be seen the small signal box which was worked by a porter signalman when required – which wasn't very often. In the siding can be seen the camping coach, a popular feature of this station for many years.

Photo: C. L. Caddy

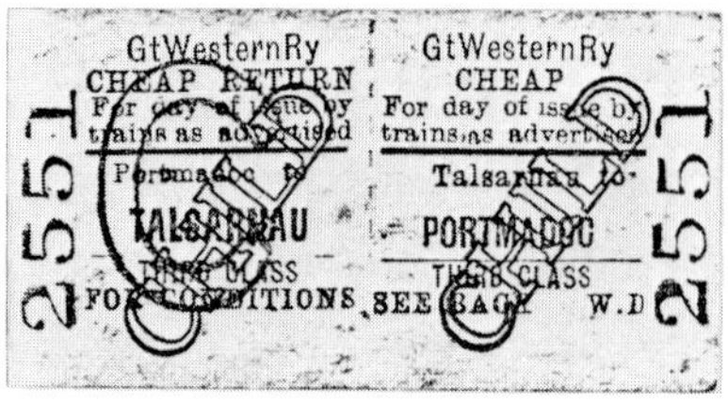

110. (Below) Llandecwyn, 25th May 1964. Situated overlooking Traeth Bach, the GWR opened Llandecwyn Halt on 18th November 1935. From the platform it is possible to see the Italianate village of Port-Meirion, location for television programmes such as 'The Prisoner', across the estuary. The houses from where the halt gets its name can be seen in the background.

Photo: C. L. Caddy

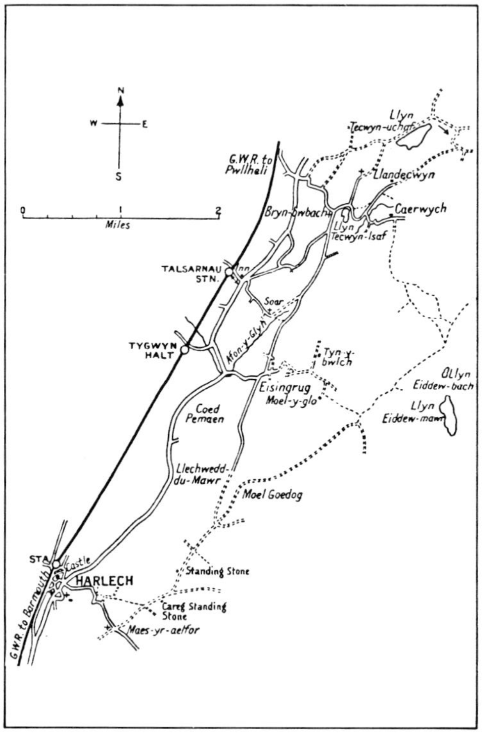

111. (Below) Llandecwyn, 26th July 1955. GWR 2-6-2T No. 4549 of the smaller 45XX class, with parallel tank top design, coasts into Llandecwyn Halt with a four coach working from Pwllheli to Machynlleth. In the background, to the left of the smokebox can be seen the toll house where road vehicles using the toll road pay their dues. An open seat sports car races the train, just visible on the right hand edge of the picture.

Photo: A. Donaldson

LLANDECWYN HALT

A journey of forty three minutes or so by stopping train from Barmouth towards Pwllheli brought the traveller to Llandecwyn Halt, a distance of 15 miles 4 chains from Barmouth. The line had climbed up a short stretch of 1 in 102 just prior to reaching Llandecwyn Halt from Talsarnau, but in fact the gradient levelled out at the Halt before resuming the climb up to Penrhyndeudraeth.

This unstaffed halt consists of a small platform, of timber construction, 75 feet in length and located on the Up side.

It came under the control of the Stationmaster at Penrhyndeudraeth. A small corrugated sheet shelter gave limited cover for passengers, the regulation lamp and nameboard providing the residue of the furniture. Access to the halt was by a footpath off the secondary toll road that crossed the estuary parallel to the railway line. The halt is still operational to this day.

112. Traeth Bach, 20th August 1966. British Railways Standard class 4 4-6-0 No. 75002 heads a four coach local passenger train bound for Barmouth and Machynlleth. Just above the second coach can be seen the timber bridge which carries the railway and the road over the waters of the estuary and the steam from the locomotive obscures the toll gate and house where the motorists are levied for using the short cut. The width of the road across the bridge is very restricted, and traffic lights control movements. *Photo: M. S. Welch*

PENRHYNDEUDRAETH

Penrhyndeudraeth and the Cambrian Coast line had an unusual beginning to its involvement with railways. Insofar as the coast line was built in sections by the Cambrian Railway, reciprocal powers between itself and the Carnarvonshire Railway (Afonwen to Caernarvon), when the Carnarvonshire Railway commenced operating in 1867, allowed for the running of their trains beyond Afonwen as far as Penrhyndeudraeth until the Cambrian Railways opened their line from Afonwen to Pwllheli. The period of operation was short lived, from September 2nd to October 10th 1867, a period of 39 days.

Penrhyn, as it was sometimes abbreviated to, is 15 miles 66 chains from Barmouth, a class B local passenger train taking Forty one minutes for the journey in 1954. The station was on level ground although from Llandecwyn Halt the gradient rose and fell at 1 in 132 for a short distance. However, beyond Penrhyn the line climbed steeply at 1 in 65/212/373 to the summit at Minffordd where freight trains in either direction stopped to pin down brakes on unfitted vehicles before descending the grade of 1 in 50 for about half a mile. This caused operational problems and piloting of freight trains over Minffordd summit in both directions was the normal practice. There were specific duties at Portmadoc shed to cover such workings and pilot locomotives were attached at Penrhyn and worked through to Portmadoc, where they were detached, and vice versa.

The establishment of Cooke's Explosives Factory at Penrhyndeudraeth provided a considerable amount of traffic over the line and was probably the mainstay of freight working until the facility was withdrawn and transferred to the Conway Valley line, carrying the products by road to the rail head where it finished its journey to its destination by rail. This traffic continues to this day. There was other freight work but this was the more normal branch line freight – livestock, general merchandise and domestic coal supplies.

The station comprised of a single passenger platform on the Up side. The Down line was a passing loop and was 936 feet in length, whereas the Up loop was 888 feet. There was a goods line off the Down side where freight workings attached/detached the pilot engine.

A siding off the platform line at the Portmadoc end facing west gave access to a coal siding and coal wharves and was locked by Annett's Key. At the south end of the Up road a line led off to another loop which gave access to the explosives factory siding and a carriage dock. On the platform, the main station building was similar in design to Talsarnau, with a cluster of corrugated sheet buildings at the Portmadoc end, and at one time a refreshment hut. At the south end of the platform was a small brick based signal box, containing the single line token instruments and an 18 lever frame. There was the electric token changing equipment apparatus for setting down the train staff at the Harlech end of the platform at the foot of the ramp and close to the signalbox.

113. (Above) Penrhyndeudraeth, 2nd October 1972. An overall view of the station and its buildings, clearly emphasising the short length of the platform. Behind the station building can be seen the village of Penrhyndeudraeth with the church just visible behind the station roof.
Photo: C. L. Caddy

114. (Centre) Penrhyndeudraeth, c. 1950. Taken from the platform and looking towards Barmouth, this view shows the signalbox, with a mineral wagon in the carriage loading siding beyond. In the distance can be seen the Cooke's Explosives Factory with wagons standing on the loop line. The public road to Llandecwyn follows the railway round in front of the factory, although the toll house is out of sight. *Photo: J. H. Moss*

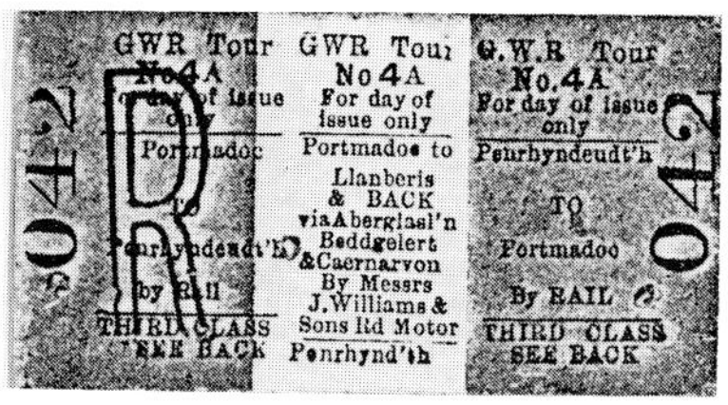
042
GWR Tour No 4A For day of issue only
Portmadoc to Penrhyndeudt'h by Rail
THIRD CLASS SEE BACK

GWR Tour No 4A For day of issue only
Portmadoc to Llanberis & BACK via Aberglasl'n Beddgelert & Caernarvon By Messrs J. Williams & Sons Rd Motor
Penrhynd'th

G.W.R Tour No. 4A For day of issue only
Penrhyndeudt'h TO Portmadoc By RAIL
THIRD CLASS SEE BACK
042

115. Penrhyndeudraeth, 1st August 1950. Former Cambrian 0-6-0 No. 892 of Portmadoc shed stands in the Down loop with a train of vans. Many of this class of locomotive survived into BR ownership but were replaced by the BR Standard class 2 2-6-0 engines. Once again, Aston's enamel signs are conspicuous on the platform. The signal box sign rests at the foot of the cabin, probably being replaced by BR Brown enamel version.
Photo: H. B. Priestley.

116. (Above) Penrhyndeudraeth, 2nd October 1972. A view from the freight loop showing the comparatively short platform. In the siding beyond the signalbox can be seen two mineral wagons in the carriage dock, and covered explosives wagons beyond. This traffic has ceased from Penrhyndeudraeth and explosives are now transported by road to Maentwrog Road station on the former Bala to Blaenau Ffestiniog branch where a new siding has been built to transport the traffic via the Conwy Valley branch.
Photo: C. L. Caddy

117. (Centre) Penrhyndeudraeth, 2nd October 1972. The signalbox at Penrhyndeudraeth housed an 18 lever frame, and was compact and attractive. The covered steps were not just ornamental for every winter, the box faced into the prevailing wind. Note the essential transport for every country signalman at the foot of the steps – the bicycle. Viewed towards Portmadoc.
Photo: C. L. Caddy

118. Penrhyndeudraeth, c. 1950. Taken from the platform looking towards Minffordd, the 1 in 65 gradient becomes apparent and a heavily loaded freight train would need every bit of assistance that the pilot locomotive could give. Descending freight trains treated the grade with the greatest respect and caution, and no doubt the thought of an explosives factory at the foot of the hill was not entirely absent from footplate crew's thoughts.
Photo: J. H. Moss

119. Minffordd, c. 1935. One of the more unusual stations on the line was at Minffordd. Constructed by the Cambrian Railways in 1873 it served as an interchange for freight and later passengers with the Festiniog Railway. The bridge over the track carries the narrow gauge line and passengers transferred along an inclined footpath connecting the two stations. The siding to the right of the picture leads into the goods yard where tranship facilities were available between the different railway company stock.
Photo: Lens of Sutton

MINFFORDD

Minffordd, at the summit of the 1 in 65/212/373 climb from Penrhyndeudraeth is 16 miles 78 chains from Barmouth, and the journey took around 45 minutes for most passenger trains. Freight workings usually picked up a pilot if the loading of the train was outside the specified limits and times varied.

The station platform was just below the summit, on the Portmadoc side of the section and comprised of a single platform on the Up side, on a curve, vision being hampered by the fact that the Festiniog Railway's line passed over the standard gauge tracks on a substantial stone bridge. There was a wooden shelter, with an office at the Portmadoc end, and a signal box with a nineteen lever frame located on the platform. The station was not a staff token exchange point but a halt on the Penrhyndeudraeth to Portmadoc section. Nevertheless when the line was constructed, there was considerable movement of slate, and the Cambrian company built a substantial goods yard with interchange tranship sidings and a warehouse. Access to the goods yard was from a facing point off the main line in the Down direction. This point was worked from a ground frame and housed in the cabin which was locked by the train token. There was an intermediate token key box provided enabling freight trains to be shunted into the yard, and following a set procedure, clear the section, but only after advising the signalman at Portmadoc that a train was safely locked in the sidings, but that the main line was now clear. Portmadoc advised the Penrhyndeudraeth signalman, and both signalmen then turned pointer switches. Up freight trains also performed any shunting and sorting in the yard, with the train brake van and freight stock for points south left pinned down on the main line. Also, at times, the work in the goods yard required a pilot loco to be sent out from Portmadoc and it was then the practice to couple this engine to a through working as far as Minffordd where it was then detached. The frame was opened, using the train staff, and the pilot loco admitted to the goods yard whereupon the staff was withdrawn from the frame, the through train proceeding on its way. The engine at Minffordd performed whatever shunting movements was required and then regained the main line later in the day, working forward to Penrhyn where it detached and worked back to Portmadoc. At other times, if there was time to spare, the pilot worked a trip turn to Penrhyn, calling at Minffordd to shunt before completing the trip to Penrhyn. Requirements were specific for working the yard and it was necessary to place wheel blocks across the rails and secure same by padlock and key before leaving. Some of the sidings were restricted to certain classes of locomotives.

The station was staffed during operating hours and a porter signalman was responsible for working the frame. Guards conducted shunting requirements.

Between Minffordd and Portmadoc was Minffordd Quarry which provided much granite traffic for the line, connection with the main line facing Portmadoc, the points being worked by a ground frame and locked with the electric train token. The main line was on a gradient of 1 in 50, falling towards Portmadoc, and care was exercised by the guards when shunting. Normally a shunting engine and van worked out from Portmadoc and it was permitted to propel wagons provided the brake van was leading and the guard was vigilant. The guard was also responsible for working the ground frame at the quarry. After work, the train returned to Portmadoc yard.

FESTINIOG RAILWAY.
NOTICE.— This Ticket is issued subject to the conditions on the Time Tables of the Company.
BLAENAU FESTINIOG
TO
PORTMADOC
FIRST CLASS Fare 2/9
4739

120. Minffordd, 13th July 1964. The gradient on the main line is clearly shown in this view, as is the curvature of the platform. The access to the goods yard leads off to the left, and the curve to the sidings is even more pronounced, which made shunting a tedious process, particularly when the loco crew were out of sight. The platform is clean and tidy albeit devoid of passengers, and in the background can be seen 'Y Garth', being eaten away by quarrying. *Photo: C. L. Caddy*

121. (Above) Minffordd, 16th September 1949. A view of the exchange sidings and the goods yard, showing the narrow gauge wagons with their load of cut slates ready to be transferred to standard gauge stock. An assortment of standard gauge mineral wagons plus three covered vans adorn the yard. Note the stacked slates by the lorry at the right hand edge of the picture. Nevertheless the scene is deceptive, for despite the wagons of slate visible, the Festiniog Railway had ceased operations in 1946, and slates were brought down to the yard by lorry. *Photo: R. S Carpenter*

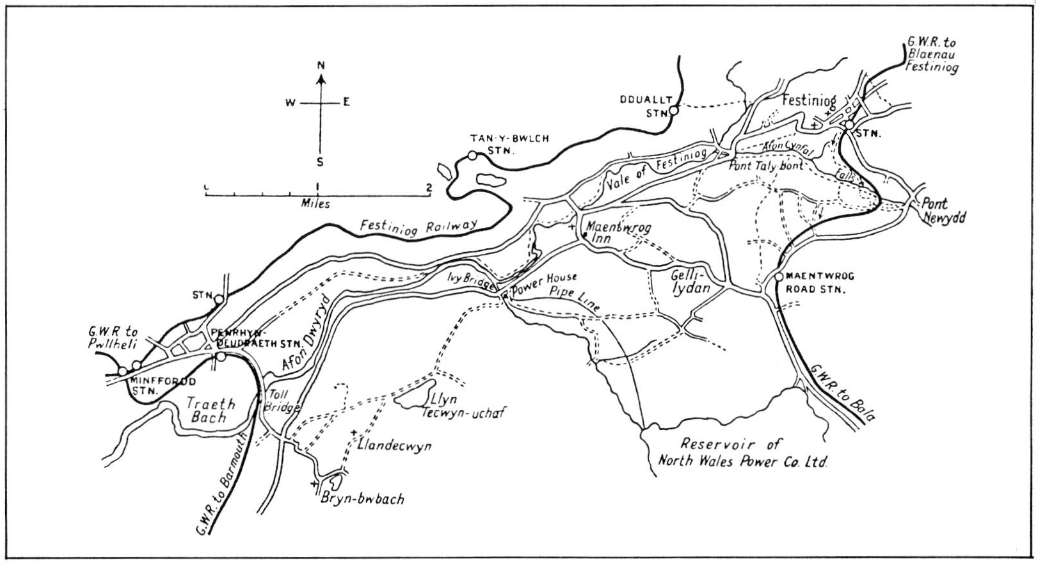

Centre: Reproduction of the map produced for those who wished to go rambling around the Vale of Festiniog via Penrhyndeudraeth, Maentwrog, Festiniog and Cynfal Falls.

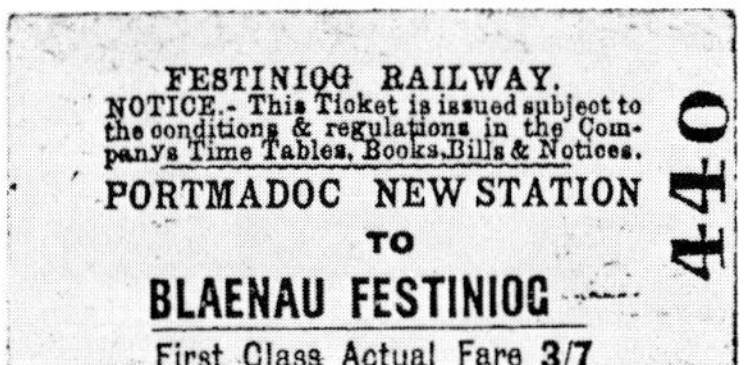
FESTINIOG RAILWAY.
NOTICE.- This Ticket is issued subject to the conditions & regulations in the Company's Time Tables, Books, Bills & Notices.
PORTMADOC NEW STATION
TO
BLAENAU FESTINIOG
First Class Actual Fare 3/7
440

122. Minffordd, 22nd April 1961. Arriving at Minffordd with Festiniog Railway Society special is "Manor" 78XX Class engine No. 7824 *Iford Manor*. An annual attraction, this special working ran from Paddington to Portmadoc for the FRS annual general meeting. Behind 7824 is "43XX" Class 2-6-0 No. 7310. Coincidental with the special, the Festiniog Railway double Fairlie locomotive *Taliesin* was renamed *Earl of Merioneth* in a special ceremony at Minffordd.
Photo: The late G. H. Platt

FESTINIOG RAILWAY

The internationally acclaimed Festiniog Railway is well documented and requires little expansion in these pages. However the need to transport slate more efficiently from the quarries at Blaenau to Portmadoc for transhipment saw the eventual birth of the Festiniog Railway in 1836. Steam haulage of public passenger trains began in 1865 with 0-4-0 side tank tender locomotives in charge, No. 1, *The Princess* and No. 2, *The Prince* being the first locomotives acquired. The railway as holiday makers and tourists know it however is a far cry from the run down concern that saw the eventual abandonment in August 1946. Fortunately, the rails remained in situ and a scheme to re-open the railway as a tourist attraction gained momentum with the result that on 23rd July 1955, passenger trains ran once more albeit the short distance from Portmadoc to Boston Lodge.

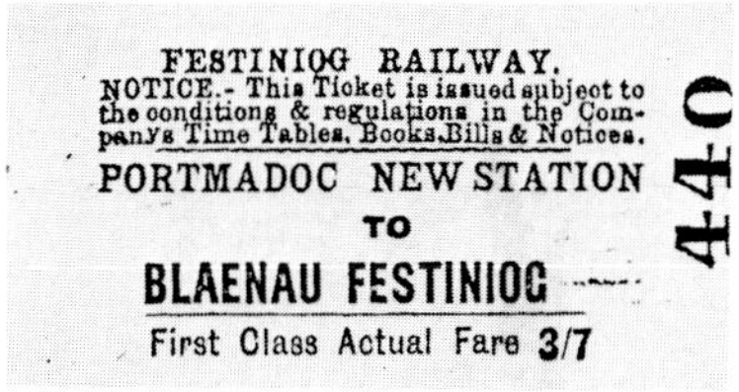

123. Portmadoc, 8th August 1935. The penultimate year of passenger services from Portmadoc over the Welsh Highland route saw examples of both company's motive power awaiting departure at Harbour station. On the left with stock for a Beddgelert train is American built 4-6-0 Baldwin locomotive No. 590. The famous "Fairle" double boilered locomotives are represented by No. 9 Taliesin, built 1876, on a train for Blaenau.
Photo: H. F. Wheeler

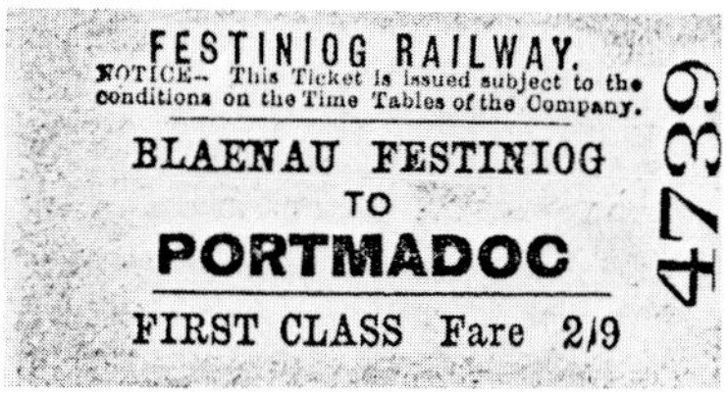

124. (Centre) Portmadoc, 24th June 1957. Much had been achieved by the Festiniog Railway Society in the short time since reopening with trains now running to Penrhyn. Here we see *Prince* at the head of a three-coach train waiting to leave Harbour station. *Photo: Norman Jones*

125. (Left-Lower) Penrhyn, 24th June 1957. Limit of working at the time, Penrhyn witnesses more activity as *Prince*, having run round its train, moves gently forward to re-couple for the return journey to Portmadoc. Services were extended to Tan-y-Bwlch in 1958, Dduallt in 1968 and Tan-y-Grisiau in 1978. Blaenau was finally reached in 1982 when a new station was built alongside British Rail's southern terminal of the line from Llandudno Junction. *Photo: Norman Jones*

WELSH HIGHLAND RAILWAY

The Welsh Highland Railway nowadays established once again in Portmadoc, briefly shared facilities with its narrow gauge neighbours, the Festiniog, at Harbour station. The Welsh Highland Light Railway came into being as the result of a merger between the North Wales Narrow Gauge and the Portmadoc, Beddgelert and South Snowdon Railways. Entry into Portmadoc along an upgraded Croesor Tramway took place on 1st June 1923 into a station on the south side of the Cambrian line which it crossed on the level. A week later, the Festiniog opened a short connection to enable both companies trains the facility of through working. Economy measures in 1929 by the Welsh Highland saw them curtail services to just north of the standard gauge crossing to avoid maintenance charges. As a result a new station, adjacent to Beddgelert Siding, opened in 1931, passengers walking the short distance to connect with Festiniog trains. The highly attractive scenery to the north did not entice customers in sufficient numbers, and so the company suspended its passenger services in December 1924, reintroducing them on a summer only basis. In 1933 the Welsh Highland once again suspended its services but a lease was taken by the Festiniog to operate the line with effect from 9th July 1934. This change did not achieve the hoped-for success and passenger trains finally ran for the last time on 26th September 1936. Freight services lasted a little longer but ceased from 19th June 1937.

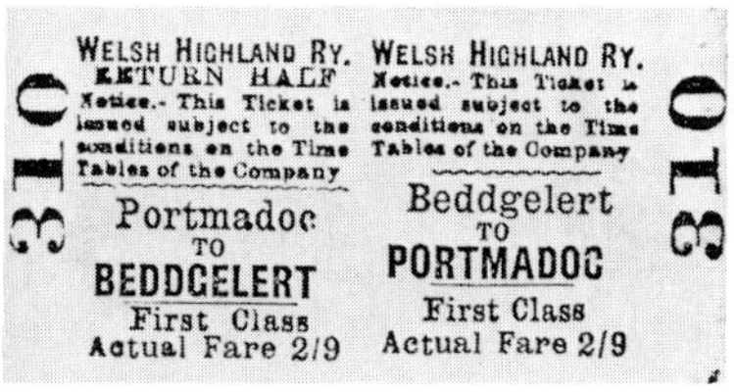

WELSH HIGHLAND RY.
RETURN HALF
Notice.- This Ticket is issued subject to the conditions on the Time Tables of the Company
Portmadoc
TO
BEDDGELERT
First Class
Actual Fare 2/9
310

WELSH HIGHLAND RY.
Notice.- This Ticket is issued subject to the conditions on the Time Tables of the Company
Beddgelert
TO
PORTMADOC
First Class
Actual Fare 2/9
310

126. (Top) Portmadoc, August 1936. This view shows the goods shed at Harbour station with the connection to the Welsh Highland going right. The vehicle to the right was a Baldwin 0-4-0 petrol engined tractor with jackshaft drive.
Photo: R. W. Kidner

127. (Centre) Portmadoc, 8th August 1935. Present day motorists would be somewhat surprised to meet a train half way along High Street. However in the summer of 1935, this view from a Beddgelert bound train hauled by Baldwin 4-6-0 side tank locomotive No. 590 shows an almost deserted thoroughfare.
Photo: H. F. Wheeler

128. (Right) Portmadoc, c. 1923. The level crossing at which point the Welsh Highland passed over the metals of the Great Western. The 1923 station is situated adjacent to the water tower behind the gates. The signal box was manned by GW men who handsignalled trains across.
Photo: Authors Collection

129. Portmadoc, June 1953. In 1953, British Railways introduced 'The Coronation Land Cruise' trains, which ran a circular route on Tuesdays, Wednesdays and Thursdays, starting at Llandudno and Rhyl and running up the Vale of Clwyd to Corwen, Western Region metals were joined. After a brief stop for water, the trains, running about half an hour apart, proceeded to Delgellau, running along the shore of Bala Lake, the instructions at this point being very specific – Do Not Exceed 25 mph. From Dolgellau the trains continued to Barmouth Junction and then to Barmouth where a break of about an hour was taken for passengers to stretch their legs and the footplate crews to perform locomotive duties. The trains then continued through Harlech to Portmadoc and Afonwen, where they rejoined the London Midland Region, and proceeded to Caernarvon where water was taken before returning home to the starting points. The excursion was an immediate success and ran in successive years, renamed as 'The Radio Land Cruise' from 1954, although in fact trains had run from Rhyl on this circular route since 1951. In 1954, the Western Region started a third 'Land Cruise' train, worked by Pwllheli men and at first with ex GWR locomotives – Collett 0-6-0 types, but working in the reverse direction. Eventually there were five 'Land Cruise' trains working in both directions each week day during the peak holiday season before the motor car caused a falling off of traffic.

Here, No. 46434 with the Rhyl train climbs out of Portmadoc heading for Criccieth, Afonwen and home. Frank Beech, of Rhyl shed, who worked the circuit for many years, sits in the driver's seat and keeps a vigilant eye on the route. The train is composed of a variety of stock, the leading coach being a brake third, from the 'Coronation' stock that toured America in 1939. An LNWR 12 wheel vehicle comes next, and the rest of the train was just as varied. *Photo: W. G. Rear*

PORTMADOC

Next to Barmouth, Portmadoc was probably the most important station on the coast line and a position it retains to this day. In steam days a class B local passenger took forty nine minutes to travel the 19 miles 2 chains from Barmouth.

After the sharp descent from Minffordd at 1 in 50 the line ran on the level for about a mile before a short rise of 1 in 410 when it levelled out again for the spring into Portmadoc. Passengers looking out from the train towards the sea could sometimes see Festiniog Railway trains crossing the embankment or 'cob' as it was locally known.

Before running in to Portmadoc, the line crossed over Traeth Mawr level crossing which was under the control of a crossing keeper. Then followed the Welsh Highland Railway crossing, which for a few brief years had a small signal box manned to control conflicting movements, all too soon becoming reduced to an 'as required' working, mainly to reduce costs to the narrow gauge line. Beddgelert Siding came into view on the Up side, where tranship siding facilities for slate traffic off the Welsh Highland were effected, and which remained as a refuge siding well after the WHR had ceased to operate. It is somewhat ironic that when British Railways decided to take up Beddgelert Siding, the Welsh Highland Railway (1964) Ltd. acquired the complex, and set up their line which is the basis of their operations today. It is pleasing to record that at long last after many years of battling against officialdom, indifference, and indeed downright opposition, the local authority have at last agreed to support the company proposal to extend their operations to Pont Croesor. Consequently, in January 1988, application was made for the formal extension of the railway. Entering the town, the main line runs over another level crossing, movements over which are controlled from the signal box on the Down platform of the station. This box has 38 levers and is operated by signalmen who are the only survivors of a once thriving station complex. The Down platform loop is 945 feet in length, whilst the Up loop is slightly shorter at 870 feet. The main buildings are on the Up platform, and at one time housed a refreshment room as well as the usual offices. A canopy extended over the platform for the entire length of the buildings. The Down platform had a wooden shelter with some enclosed accommodation and a canopy covering the platform offered passengers some protection. Beyond the buildings at the Criccieth end an open footbridge, subsequently removed, provided access between the platforms. Water columns were located on both platforms. There was an auxiliary electric train token instrument located at the Criccieth end of the Down platform which enabled the signalman in the box by the level crossing to release a token for the Portmadoc to Criccieth section – without the need to walk the entire length of the platform – when a Down train had to wait for the section to be cleared.

An extensive goods yard on the Down side provided storage for freight stock and a small goods shed adjoined the platform. There was also siding accommodation for passenger coaches within the yard. It was the practice for Portmadoc to be the starting and terminating point for most freight trains in either direction, and almost every freight working was re-manned here.

On the Up side at the Criccieth end of the station, a two road motive power shed was provided. In GWR days it was coded PMDC and in 1947 a total of nine locomotives were allocated there although it was regarded as a sub shed to Machynlleth. It was the last stronghold of the ex Cambrian Railway 0-6-0 and members of the outside framed 4-4-0 90XX class were regularly attached to the shed. There were fourteen sets of men based here for many years although this diminished towards the end and on closure of the shed the staff and remaining work transferred to Pwllheli. Some staff retired rather than move along the coast. Work from the shed included pilot duties assisting trains over Minffordd, freight workings to Barmouth and Pwllheli, as well as passenger workings between Pwllheli and Machynlleth. According to Mr. Starling former Portmadoc driver, Portmadoc men learn the road from Dovey Junction to Aberystwyth, an operation which proved abortive. After nationalization it was proposed that they even learned the road to Caernarvon and Bangor but in the event, regional boundaries prevailed and the proposal came to nothing although it is understood that just prior to WW2, excursion traffic from Chwilog to Shrewsbury via Afonwen and Machynlleth was covered by Portmadoc men who worked from Chwilog without troubling the LMS for the provision of a locomotive! When nationalization materialised however, one immediate effect was to eliminate some of the work at Portmadoc and Pwllheli sheds, which previously had required the working of through coaches off the Caernarvon line to destination, whereas Bangor men worked their own coaches to Pwllheli, thus avoiding light mileage to and from Afonwen. Indeed the practice had commenced before the 1948 Act, when 'The Welshman' through coaches from Portmadoc to Euston were worked by Bangor men from Portmadoc. A Stanier class 3 2-6-2T performed this duty each summer for seventeen years. The routine was for the 2-6-2T to light engine from Bangor to Portmadoc and extract two corridor coaches from the sidings. The train then proceeded to Afonwen where two coaches from Pwllheli were attached and the combined train worked forward to Bangor where the tank engine came off and was replaced by a tender locomotive. The powers that be decided however that once a day was enough for the intruders and the afternoon return working was under the control of the GWR/Western Region, a pattern that persisted until the working finally ceased. In the busiest weeks of the year, for about four weeks, it was not unknown for the coach sets to be doubled, with four each from Portmadoc and Pwllheli, although in these cases, extra motive power was available from Afonwen to Bangor, in the form of 2-6-4T which was worked light engine to Afonwen, being prohibited from the section to Portmadoc. In the LM Region RSD notice No. 15 for Saturday, 11th August 1951, the down 'Welshman' (11.15 a.m. ex Euston) shows the working as follows:

Clause 340. Corridor Stock (4) and usual circuit (4) including RF Euston to Bangor, with usual Portmadoc and Pwllheli-Euston (7). (Total 15 vehicles). This indicates one 3 coach set and one 4 coach set working beyond Bangor to Afonwen. The 3 coach set would work to Portmadoc.

The district normally held spare coaches in reserve during the week, Portmadoc holding a brake composite coach and Pwllheli two third class coaches. On Saturdays, in the summer, these would be worked forward to Chester, strengthening a five coach formation from Pwllheli (Circuit 27). A balancing set would be worked on the Saturday, with three extra vehicles, which replaced the spare coaches at Portmadoc and Pwllheli. The five coach set worked a five day circuit during the week and repeated the process every Saturday.

130. Portmadoc, 8th September 1953. A small hill east of Portmadoc station provides a fine vantage point to watch trains. Here 'Dukedog' 4-4-0 No. 9000 leaves Portmadoc with an afternoon working for Dovey Junction. Behind the last coach is the gate-keeper's cottage at the former Welsh Highland Railway crossing where the connection to the Festiniog Railway crossed the standard gauge. Behind the centre coach can be seen Gelert Farm, now the workshops of the Welsh Highland Railway (1964) Ltd. and the formation of the former WHR trackbed running at an angle to the BR line. Behind the fence in the middle of the photograph and buried in the grass lies Beddgelert Siding. Ysgol Eifionydd hides under the shade of trees.
Photo: W. G. Rear

131. Portmadoc, c. 1957. Displaying the Cambrian Coast Express headboard, and carrying Class A lamps, British Railway Standard class 2 2-6-0 No. 78000 runs across the level crossing and into Portmadoc station with the Pwllheli portion of the train from Paddington. Beyond the level crossing, the Up and Down main line join via the crossover at the commencement of the single line to Minffordd. The Up line continues towards Beddgelert Siding and the lower signal arm controls this line. Underneath the signal can be seen freight stock, standing in the cattle dock.
Photo: C. L. Caddy Collection

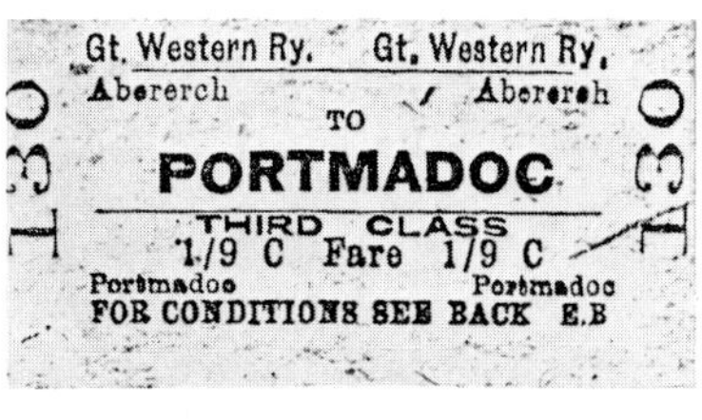

Gt. Western Ry. Gt. Western Ry.
Abererch / Abererch
TO
PORTMADOC
THIRD CLASS
1/9 C Fare 1/9 C
Portmadoc Portmadoc
FOR CONDITIONS SEE BACK E.B

132. Portmadoc, 19th August 1958. 'Duke-dog' 4-4-0 No. 9021 draws a freight train, composed mostly of vans, across the level crossing towards Minffordd. The fireman looks back to check that all is well. A brake van rests in the Down platform.

Photo: A. Donaldson

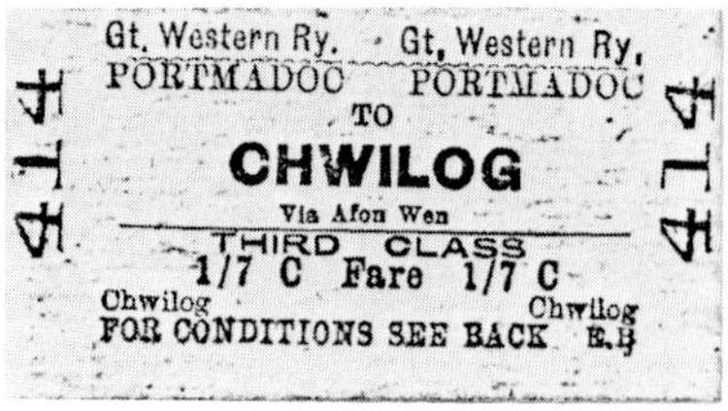

Gt. Western Ry. Gt. Western Ry.
PORTMADOC PORTMADOC
TO
CHWILOG
Via Afon Wen
THIRD CLASS
1/7 C Fare 1/7 C
Chwilog Chwilog
FOR CONDITIONS SEE BACK E.B
414 414

133. (Centre) Portmadoc, 13th July 1964. A view from the Down platform looking towards Pwllheli showing the substantial station buildings on the Up platform. The wooden shelter on the Down platform is of Cambrian origin whilst beyond is the goods yard with an abundance of stock visible. Notice the large water tank just visible behind the station buildings.

Photo: C. L. Caddy

134. (Below) Portmadoc, 20th August 1966. It was the practice for trains to take water at Portmadoc, prior to the haul over Minffordd and Barmouth. Here, the fireman tops up the tank of BR Standard Class 4 4-6-0 No. 75009. The level crossing gates were closed across the tracks to clear the road congestion during such operations. This working was the Saturdays Only from Pwllheli to Birmingham, only two more Saturdays remaining for regular steam working on the Cambrian Coast before such workings disappeared from the working timetable until the return to steam occurred in 1987.

Photo: M. S. Welch

135. Portmadoc, c. 1958. Ex GWR 2-6-2T No. 4560, based for many years at Portmadoc, heads out of the Down platform with a lightweight freight for Pwllheli. The tank engine is a member of the earlier class of locomotives, with straight sided tanks. Note the wooden post for the lower quadrant starter signal, behind which is the small shelter where train crews could release a single line staff for the section to Criccieth, thus avoiding a long trek to the other end of the platform to the signalbox. The goods shed is on the right hand side of the picture. *Photo: A. Donaldson*

136. (Centre) Portmadoc, c. 1950. At the Pwllheli end of the Up platform was situated a two road locomotive shed which was in fact a sub-shed to Machynlleth, sharing the work along the coast line with Pwllheli and the parent shed. The original building was a wooden structure, thought to have been dismantled and re-located at Pwllheli in 1907, when the later structure appeared. The shed closed in August 1963 and unfortunately has now disappeared. The allocation here comprised mainly of 0-6-0 tender engines, of various vintages, for about half the work of the shed was freight traffic. A couple of 2-6-2T engines of the 45XX or the larger 4575 class were normally based here. There was no turntable at Portmadoc so engines turned at Afon-wen until that facility was removed when it became necessary to go to Pwllheli to turn. *Photo: J. H. Moss*

137. (Below) Portmadoc, c. 1954. Prairie tanks Nos. 5570 and 4575 stand resting on shed and facing the correct direction for their next working. Both engines were normally resident at Portmadoc for many years. The shed was of brick although the original slate roof was replaced with corrugated asbestos sheeting a couple of years earlier. The building measured 120 feet by 35 feet, including the small offices attached to the rear gable. The proximity of housing to the shed ensured that foot-plate crews were extremely careful about smoke emission and noise although it was accepted as normal in those 'bad old days'. *Photo: M. S. Welch*

138. (Left) Portmadoc, 19th August 1958. 'Dukedog' 4-4-0 No. 9021 engaged in shunting operations, pulls a 'toad' brake-van along the Up platform line. The Dukedog displays the 89C shedplate and the boiler has a separate top feed which was not found on every member of this class. *Photo: A. Donaldson*

139. (Left-centre) Portmadoc, c. 1953. Cambrian 0-6-0 No. 894, still bearing GWR on the tender sides, stands at the down platform with a freight working. The water 'bag' is in the tender, topping up the tank, whilst the crew on the footplate attend to their duties. Note the tarpaulin draped over the cab roof – a vital piece of equipment when the weather turned nasty and tender-first running was unavoidable. These locomotives were the remaining survivors from the pre-grouping days and were powerful machines. However, rough riding became evident when they were due in the workshops and as such were prone to roll alarmingly on some sections of track. Originally built by Beyer Peacock & Co. in Manchester in 1908, this example survived the rebuilding and 'Swindonising' required by the GWR until finally withdrawn in 1954, never actually carrying the Lion & Wheel emblem on its tender. *Photo: M. S. Welch*

141. (Below) Portmadoc, c. 1954. An interesting view of 9005 taken from the footbridge at Portmadoc and looking towards Pwllheli. The somewhat exposed conditions on the footplate are clearly illustrated here, and the tarpaulin had to be accessible. This locomotive had no separate top feed but otherwise it was typical of the class. The driver does not appear to have a seat although the position of the screw reverser didn't give much space on the right hand side and so most drivers performed their duties standing. Unless you were tall, it was difficult to see out of the spectacle plate windows. Here, the starter had 'dropped' and the crew were pulling away from rest on their journey to Afonwen. *Photo: M. S. Welch*

140. (Left) Portmadoc, c. 1953. Following nationalization, the London Midland Region commenced running through to Portmadoc although workings were restricted to the Up 'Welshman'. Here No. 40132, a Stanier 2-6-2T from Bangor, tops up the tank prior to departure. It would attach the coaches from Pwllheli at Afonwen and then work both portions forward to Bangor where a tender engine would work forward the stock to Chester and Euston. In the goods yard some vans are being emptied. Note also the horse drawn cart.
Photo: M. S. Welch

142. (Above) Portmadoc, c. 1954. Engine No. 5570 with a train from Pwllheli to Machynlleth runs past the locomotive shed and pulls into the Up platform with a four coach train of pure GWR stock. Apart from the number on the smokebox, this picture could have been taken at any time over the previous twenty years.

At one time there were two signal boxes at Portmadoc, but the West box closed in 1932 and control of points and signals transferred to the East box. The small hut at the end of the Down platform, seen here alongside the water column and the starter signal, house a small frame and token release equipment. *Photo: M. S. Welch*

143. (Centre) Portmadoc, c. 1954. British Railways Standard class 2 2-6-0 No. 78003, which was attached to Portmadoc shed, sets back from the Down platform, possibly banking a train to Afonwen over the Wern gradient. It was decreed in the Appendix to the Service Timetables that assistance could be provided at the rear of trains over the climb, the banking locomotive remaining coupled as far as Criccieth. It was somewhat unusual for locomotives to perform this duty tender first and it is possible that there was nothing available which faced the right way. Note that the signal is lowered for the main line. On the Up side of the track, the small buildings adjoining the loco shed are worth a second glance.

Photo: M. S. Welch

144. Portmadoc, 20th August 1966. British Railways Standard class 4 4-6-0 No. 75009, minus shedplate, draws into Portmadoc station with the Saturdays Only working to Shrewsbury and Birmingham. The canopy over the platform on the Down side has been removed as has the shelter, replaced by the 'Bus Stop' type. Some concession to modernisation.

Photo: M. S. Welch

145. Wern, c. 1953. From the summit of the line at Wern Siding, looking back towards Portmadoc, Collett 0-6-0 No. 2201 working the 1.35 p.m. Saturdays Only from Machynlleth to Pwllheli coasts over the summit after working hard up the 1 in 50 gradient. Driver D. Price of Machynlleth shed was at the controls but didn't look over the side until after this picture was taken. *Photo: W. G. Rear*

WERN SIDING

For about half a mile beyond Portmadoc towards Criccieth, the line remained on the level. It then commenced climbing, first at 1 in 84 and then at 1 in 50 – just as steep as the climb towards Minffordd but less well known.

Wern Siding was created to tap a supply of stone for the Cambrian Railway's own internal use remaining in use and being listed in the September 1964 working timetable.

The siding was connected to the running line facing from the direction of Portmadoc and the points were worked by a ground frame locked by the Electric Train Token. Since the line was on a 1 in 54 gradient falling towards Criccieth, it was necessary for guards to pin down brakes and secure the train before detaching the locomotive. It was also permitted to propel a brake van and wagons from Portmadoc, subject to the usual conditions for this type of working. The responsibility for working the siding lay with the guard of trains booked to call.

The siding was one mile 59 chains from Portmadoc and just beyond, the line levelled out briefly before descending towards Portmadoc. Guards of freight trains travelling towards Criccieth were required to apply their hand-brake and keep couplings tight until the bottom of the incline. If necessary, freight trains stopped and pinned down brakes, with a notice to that effect displayed on the lineside.

146. Wern, c. 1953. This small goods depot, west of Portmadoc and adjacent to the Criccieth road, existed to serve the nearby Wern estate as well as a small quarry for the railway's own use. Great Western built 2-6-2T No. 4575 leans to the curve as it passes the site of the siding entrance. A ground frame, in front of the lineside hut, controlled access to the siding. It also marked the limit of the respective permanent way gangs sections of track. *Photo: W. G. Rear*

BLACK ROCK HALT

After a distance of 22 miles 68 chains from Barmouth, Down trains reached Black Rock Halt, the last unstaffed halt on the line before economies of the late 1960s saw a reduction in overall facilities. The simple all timber platform was one hundred feet long and located on the Down side.

Originally there was a shelter, but this had and the sole item of furniture remaining was a name-board. The halt came under the control of the Stationmaster at Criccieth and access was by public footpath from the beach.

The halt last appeared in the working timetable for May 1977 having been closed to the public on 13th August 1976 for safety reasons. Official closure took place on 27th June 1977.

147. Black Rock, 3rd July 1969. To the east of Criccieth is a stretch of coastline and beach known as Black Rock sands, which over the years has grown in popularity with holiday-makers. The Great Western Railway, ever mindful of their image as a 'holiday line' constructed a small halt here and opened it to the public on 9th September 1923, serving the beach.

In this view, looking towards Portmadoc, a British Rail "corporate image" sign has appeared on the platform, keeping company with the older traditional cast iron warning signs.
Photo: C. L. Caddy

148. Black Rock, 15th July 1965. Black Rock Halt, viewed from track level over a cattle creep, shows the wooden platform with the solitary name-board. In the distance can be seen the outline of Criccieth Castle, which dominates the local skyline.
Photo: C. L. Caddy

149. Black Rock, September 1953. The line ran alongside the shingle and beach and was somewhat exposed to the elements blowing off Cardigan Bay. Here, between Black Rock Halt and Criccieth, 2-6-2T No. 5570, from Portmadoc shed, works the 10.50 a.m. freight from Barmouth to Pwllheli East coasts gently despite the ascending grade. The working had been re-manned at Portmadoc, and the crew had another hour to pass before they arrived at Pwllheli. No need to rush! *Photo: W. G. Rear*

150. Criccieth, September 1953. The ex-Cambrian Railways 0-6-0 tender engines were still much in evidence on the coast line, and here, 894 works the 3.45 p.m. Pwllheli to Dovey Junction passenger out of Criccieth towards Portmadoc. Beyond the road bridge is Merllyn Crossing, protected by signals interlocked with the gates, hence the 'clear' indication for the opposite direction.
Photo: W. G. Rear

CRICCIETH

Approaching Criccieth from Portmadoc, the line crosses two public roads by means of level crossings. The first, nearer Portmadoc, was known as Merllyn level crossing and the keeper tended the crossing gates manually until recently. He had warning bells to advise him of impending train departures from both Portmadoc and Criccieth and there were fixed time limits for the crossing which was normally open to road traffic and against the railway to close the gates.

The second, known as Maes level crossing, was located in the centre of Criccieth. The gates were again manually operated and normally left across the railway, and a crossing keeper lived alongside.

Criccieth station is 24 miles 14 chains from Barmouth and local stopping trains were allowed sixty three minutes journey time. The station was a crossing point, with the Down loop at 777 feet in length, the Up loop being slightly shorter, at 771 feet. The line climbed at a gradient of 1 in 82 to reach the station, and following a short level stretch, fell away again at 1 in 84 towards Afonwen.

If trains were exceeding their loading limits, assisting engines could be called out from Portmadoc to Criccieth to provide banking assistance. Locomotives could be used to bank trains from the rear between Criccieth and Portmadoc in either direction but if the engine was required to provide assistance beyond Portmadoc then instructions specified that the assisting engine was coupled to the train engine. Because of the gradient, the usual precautions were required to be taken by guards when dealing with Up trains. The main buildings were located on the Up platform, the canopy having been part of a package of improvements carried out by the GWR. There was also a short bay/loading dock at the Portmadoc end of the station on the Up side.

On the Down platform, there was a canopy which extended the same length as that on the Up platform although the waiting facilities were less generous. The structure, which had some enclosed space, was constructed of wood. A siding ran behind the Down platform for about half its length, connection being off the Down loop. A signal box which housed 22 levers was located at the Portmadoc end of the Down platform and a public pedestrian crossing with gate locks worked from the cabin stood at the east end at the foot of the platform ramps. Also on the Up side was a goods warehouse and two sidings.

Criccieth lost its crossing facilities with the 1968 issue of the working timetable. The bay lines were removed some time before this. Goods facilities had been withdrawn from the 5th October 1964.

WESTERN REGION

Day Excursion Bookings

EACH WEEKDAY

Commencing 12th June, 1961 and until further notice

TO

PORTMADOC, CRICCIETH
and PWLLHELI

FROM	DEPART	RETURN FARES (Second Class)		
		Portmadoc	Criccieth	Pwllheli
	a.m.	s. d.	s. d.	s. d.
MACHYNLLETH	8 5	12/6	13/9	16/-
PENHELIG HALT	8 34	10/-	11/6	13/6
ABERDOVEY	8 38	9/9	11/-	13/3
TOWYN	8 47	8/9	10/3	12/3
TONFANAU	8 52	8/-	9/6	11/9
LLWYNGWRIL	9 3	7/-	8/3	10/6
FAIRBOURNE	9 10	6/3	7/9	9/9
Arrival Times		a.m. 10 23	a.m. 10 33	a.m. 10 57

Passengers return by any Train same day affording a Service through to destination.

Children under Three years of age, Free; Three and under Fourteen years of age, Half-fare.
(Fractions of 1d. charged as 1d.)

IT WOULD ASSIST THE RAILWAYS IN PROVIDING ADEQUATE ACCOMMODATION IF INTENDING PASSENGERS WOULD OBTAIN THEIR TICKETS IN ADVANCE

TICKETS CAN BE OBTAINED IN ADVANCE AT BOOKING STATIONS.

DAY OUTINGS——PARTIES OF 8 ADULTS OR MORE AT REDUCED FARES BY PRIOR ARRANGEMENT.

Further information will be supplied on application to the Stations or to Mr. O. VELTOM, District Traffic Superintendent, Shrewsbury (Telephone Shrewsbury 3614, Extension 42); or Mr. W. R. STEVENS, Divisional Traffic Manager, Cardiff.

Paddington Station, W.2.
April, 1961.

R. HAMMOND,
General Manager.

No. 79 R.P. 12

Printed in Great Britain by G. R. Griffith Ltd., Chester

151. Criccieth, September 1953. Dukedog 4-4-0 No. 9024, with the 4.25 p.m. two coach local working from Afonwen to Portmadoc, after working a Down school children duty from Ysgol Eifionydd at Portmadoc, scampers over Merllyn Crossing with steam on, despite the falling 1 in 82 grade. The behaviour of the children was inclined to be boisterous, and guards made the plea to the footplate crew 'not to hang about'. No one ever did! Nor did they 'hang about' on the return journey, for the sooner they were on shed, the sooner they finished and went home! *Photo: W. G. Rear*

152. Criccieth, September 1953. Dukedog 4-4-0 No. 9023 runs past the goods shed with the Dovey Junction to Pwllheli working. In the foreground is the public crossing at the foot of the platform ramps. The goods shed seems closed for the day although it must be admitted that freight traffic was not the main source of revenue. Note the unusual track layout giving direct access to the goods shed from the Down loop. *Photo: W. G. Rear*

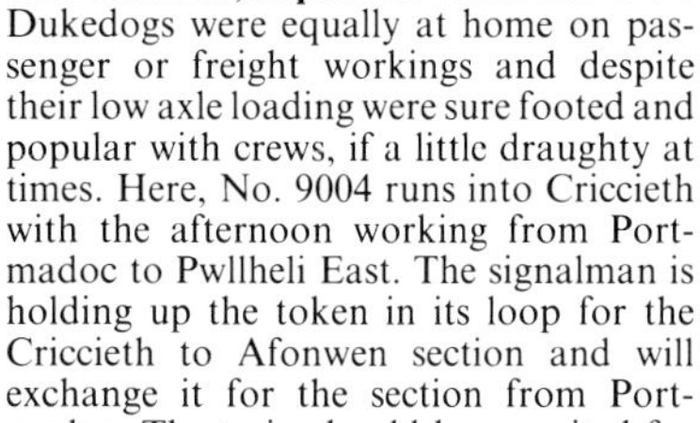

153. Criccieth, September 1953. The 4-4-0 Dukedogs were equally at home on passenger or freight workings and despite their low axle loading were sure footed and popular with crews, if a little draughty at times. Here, No. 9004 runs into Criccieth with the afternoon working from Portmadoc to Pwllheli East. The signalman is holding up the token in its loop for the Criccieth to Afonwen section and will exchange it for the section from Portmadoc. The train should have waited for ten minutes but was running late, and consequently by not having to stop, picked up time and was only three minutes down at Afon Wen where some of the load was detached for the Caernarvon line. In Cambrian days it was the policy to keep banking engines at Criccieth to assist heavy trains over the Wern to Portmadoc section. *Photo: W. G. Rear*

154. (Above) Criccieth, 3rd October 1964. A fine study of the 22 lever signalbox at Criccieth, located at the Portmadoc end of the down platform. Someone has been generous with the white paint used for marking the platform edge! Visible in the corners of the box are the token instruments, whilst the signalman's transport rests on the railings awaiting its next turn of duty. Notice the ruined castle in the background. *Photo: M. S. Welch*

155. (Centre) Criccieth, 3rd October 1964. The down 'Cambrian Coast Express', albeit without headboard, and drawn by British Railways Standard class 4 4-6-0 No. 75004, pauses at the platform before resuming the journey to Afon Wen and Pwllheli. These engines were popular with Cambrian train crews, having been brought up on the more open cab types. They were more powerful than their predecessors and the comment was made that they had arrived about ten years too late was echoed by all.

Photo: M. S. Welch

156. (Left) Criccieth, September 1947. The buildings at Criccieth viewed in the closing months of GWR ownership. Despite the shortage of passengers, there is an air of pride in the presentation of the station. The matching awnings are of unusual design, but not out of place.

Photo: Authors Collection

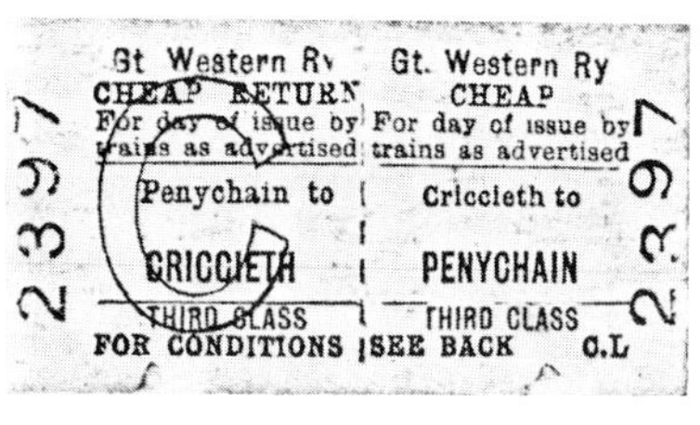

157. Afon Wen, September 1953. The 1.35 p.m. Dovey Junction to Pwllheli working, headed by Collett 0-6-0 No. 3201, runs in to the Down platform at Afon Wen. On the back platform road, 2-6-2T No. 4575 stands at the head of a one coach working from Pwllheli to Portmadoc. The Bangor men will depart at 3.50 with four coaches and a van which can be seen next to the WR composite brake coach. The Portmadoc train will wait until 4.25 p.m. before continuing. *Photo: W. G. Rear*

AFON WEN

Afon Wen station formed the junction with the LNW/LMS line from Caernarvon and was the sole reason for locating a station there in the first place. Built by the Cambrian Railways, it was a convenient point to make the connection, the interchange of traffic justifying this existence. Although the LMS worked its traffic to Afon Wen, the station was operated by GWR staff until the Cambrian network was transferred to the London Midland Region, the former dictating the operation of the station.

The station was 27 miles 47 chains from Barmouth, local trains taking seventy one minutes for the journey. The station was on level ground but only a few feet above sea level, and in bad weather it was not unknown for sea spray to be blown over trains. The track had descended from Criccieth, although there were three sections of rising gradient in the three mile stretch, one length of fifty six chains falling at 1 in 64, changing to a rising gradient of 1 in 65 for about the same distance.

The station was over half a mile in length from one extreme to the other. The main station buildings were on an island platform, on the Up side, and which contained the usual offices and a separate licenced refreshment room which provided lubrication to many a hot train crew. The Down platform had a rudimentary shelter with its back to the sea. A footbridge at the Portmadoc end enabled passengers to cross the line, but the open aspect occasionally made it an uncomfortable experience. Access to the platforms from a private road was also by footbridge at the Pwllheli end of the station. A signalbox was located on the Down platform, the frame housing 75 levers. The back platform line had a Track circuit installed, the 'Vehicle on Line' switch being located on the wall of the station building.

The Up platform loop was 1233 feet long and could accommodate the longest trains, whilst the Down loop was 1044 feet in length. There was a refuge siding, known as 'Machine' Siding on the Down side which could house 50 wagons as well as the engine and brake van. A crossover road was located near the signal cabin. A turntable was installed on the station site behind the Down platform, but this was removed in the 1930s. Two water tanks were installed on the Turntable road, also known as the 'Ash' siding which remained until the end of steam and there was also a water column at the Pwllheli end of the island platform. A terrace of employees cottages were located on the seaward side and the Stationmaster had a bungalow on the Up side which today provides the only remaining evidence that there ever was a station at Afon Wen. Also on the Up side was the 'Goods' loop which led to the interchange sidings, the practice being for LMS freight trains to work into these sidings and collect wagons for the return working at the same time. The sorting and shunting was the responsibility of the Western men and the daily workings between Pwllheli and Portmadoc covered this duty. The LMS ran three freight trains daily to Afon Wen, but only two back to Bangor, the unbalanced duty being a mixed train in the evening. This was reduced to two trains daily after nationalization.

Originally, the line to Pwllheli was single but in 1947 the section to Penychain was doubled and worked by Absolute Block. Lines to Caernarvon and to Criccieth were single and worked by train staff. The LMS line had a standard Webb pattern large token for the short run to Chwilog whereas standard GWR equipment was installed for the Afon Wen-Criccieth section.

In the 1950s traffic was at its peak on summer Saturdays and it was commonplace for one hundred plus train movements to take place between 5.00 a.m. and 10.00 p.m. Trains for Butlins off the LM Region were generally ten coach formations although eleven coach workings were not unknown. The passing loops to Caernarvon could normally only accommodate ten vehicles plus locomotives. Motive power for most of these workings were permutations of tank and tender engines, and working in pairs. Largest engines were class 5 4-6-0s. Normal motive power for the LM line in BR days were the 2-6-4T engine, of Fowler, Stanier, Fairburn and BR Standard classes. The Western Region classes have already been described or can be deduced from the text elsewhere.

158. Afon Wen, September 1953. Portmadoc depot 2-6-2T No. 4575 runs into Afon Wen with the 11.00 a.m. Ruabon to Pwllheli working. After turning there, it will return with the 3.45 p.m. to Dovey Junction as far as Afon Wen where it will pick up an additional coach, brought on the LM Region working from Pwllheli, and after a short wait will continue on to Portmadoc at 4.25 p.m. *Photo: W. G. Rear*

159. (Above) Afon Wen, c 1954. The east end of Afon Wen station looking towards Criccieth. The London Midland Region afternoon freight from Caernarvon has arrived and the Fowler 2-6-4T is shunting the 'Machine' siding collecting its stock for the return journey. In the back platform road, just visible under the footbridge, can be seen the stock which has arrived from Pwllheli and will constitute the 3.50 p.m. to Bangor. This working contained a single coach, detached here and loads which would be worked forward to Portmadoc by WR men, seen here departing off the platform. Note the footbridge connecting the Up and Down platforms. *Photo: J. H. Moss*

160. (Centre) Afon Wen, August 1960. Taken from the footbridge which connected the Up and Down platforms, looking towards Pwllheli. Passengers are moving on the Down platform and a Stanier 2-6-4T from Bangor shed stands at the platform end before moving off to the Up line. In the distance a train for Caernarvon can just be seen commencing the climb to Chwilog. *Photo: G. H. Platt*

161. (Left) Afon Wen, c. 1963. The 4-4-0 Dukedogs were replaced by British Railways Standard class 3 2-6-2T engines of the 82XXX class. They were compact and comfortable to work, and became an instant hit with most of the crews who handled them although one or two die hard veterans claimed that the Cambrian 0-6-0 tender engines were better. Nos 82031/2/3 were allocated to Machynlleth and worked exclusively over the coast line on passenger and freight duties. They were painted lined green and and were kept as clean as conditions permitted. Here No. 82033 waits at the Down platform, outside the signalbox, with a freight for Pwllheli East. The signalbox, which housed 75 levers in the frame, was the largest on the coast. *Photo: C. L. Caddy*

162. Afon Wen, 1957. 'The Welsh Chieftain' was a successor to the Land Cruise trains which operated in the early 1950s on the Rhyl, Denbigh, Corwen, Barmouth, Afon Wen, Caernarvon, Llandudno Junction, Rhyl circuit. In 1955, there were five trains making the circuit during the week daily, three in one direction and two in reverse. In 1957, the numbers were reduced, partly due to falling off of demand and the trains acquired new titles in a bid to drum up support. The 75XXX class were used extensively on all workings. This working originated from Conway and worked the circuit anti-clockwise. No. 75054 was at Rhyl and carries the 6K shedplate of that depot. Just visible at the right hand edge of the picture can be seen the starter signal for the Up platform with the route indicator box just visible.
Photo: C. L. Caddy Collection

163. (Centre) Afon Wen, September 1953. The GWR 2-6-2T engines were very versatile engines, being equally at home on passenger or freight duties and were put to good use on the Cambrian lines. Here No. 5524 from Portmadoc shed works the afternoon freight from Portmadoc to Pwllheli East. The line in the foreground is the 'Ash' siding, formerly known as the Turntable road. Here in days gone by, the station pilot awaited its time, although the particular duty disappeared many years before this photograph was taken. The two water tanks meant that two locomotives could top up their tanks at the same time, very necessary when perhaps eight engines would be requiring water on a busy summer Saturday morning. The siding was exposed to the elements, and strong wind blowing off the sea could whip sand and clinker dust into a luckless fireman's face as he struggled to keep his balance whilst filling the tanks. The driver could take shelter in the cab unless he needed to oil around his charge.
Photo: W. G. Rear

164. Afon Wen, June 1953. Until they were withdrawn, the ex-Cambrian 0-6-0 tender engines worked passenger or freight trains with impunity. They were strong engines although somewhat rough and would roll alarmingly when they were run down. Swindon tried to cure this by drilling holes in the balance weights but without much success. Here, 849, shortly to be withdrawn, pulls into the back platform road with the 10.25 a.m. Pwllheli to Dovey Junction turn. This was worked by Machynlleth men who had travelled to Pwllheli with the 6.50 a.m. early morning mail. They had thirty five minutes to turn and clean the fire before returning South. The track layout west of the platforms can be seen, and the LM line to Chwilog can be made out to the right of the engine, just above the lamp bracket. *Photo: W. G. Rear*

165. Penychain, August 1960. In happier times when steam reigned supreme, a Bangor Stanier class 3 2-6-2T No. 40132 stands at the head of a train of LNER and British Railways stock. Albert Victor Williams looks at the gardens beyond the fence whilst awaiting departure time. The class 3 tank will only work unaided as far as Afon Wen where it will run around the stock and couple to the front of a class 5 4-6-0 which will work the train to Warrington. In the Down platform another train waits, also bound for Afon Wen and the LM line through Caernarvon, this working destined for Stoke on Trent.
Photo: G. H. Platt

166. Penychain, c. 1960. Taken from the broad/Up platform face looking towards Pwllheli and emphasising the width of the platforms necessary to accommodate the volume of holiday makers who arrived at Butlins Camp by train. The name-board, ever modest, proclaims "Penychain for Pwllheli Holiday Camp" – and your holiday begins here!
Photo: Lens of Sutton

167. Penychain, June 1965. Taken from the footbridge that passed over the line at the Pwllheli end of the station, this view looks towards Pwllheli and shows the commencement of the single line. In the small hut in the foreground an electric token release instrument was located which enabled train crews who had been waiting for a staff to authorise travelling over the single line to receive a token without the need to walk back to the box. The two starters indicate that either line was signalled for two way working. *Photo: W. G. Rear*

168. Penychain, June 1965. Taken from the same location as the previous photograph but looking towards Afon Wen. The simple lines of the station buildings are clearly shown. *Photo: W. G. Rear*

PENYCHAIN

The Great Western Railway opened the halt at Penychain on 31st July 1933 in a bid to drum up traffic for their holiday resort lines. The building of a holiday camp commenced before the second world war but was incomplete at the commencement of hostilities in September 1939. The site was requisitioned by the military and became a Naval Training Base known as HMS Glendower for the duration. Special trains carrying naval personnel ran every weekend via Caernarvon to Chester and were heavily loaded without exception. After the war, the camp reverted to civilian use and opened for holiday-makers in 1947. Simultaneously being up-graded to station status although it was not staffed in the winter months. The GWR doubled the line between Afon Wen and Penchain, worked under Absolute Block Regulations. At the same time, a second platform face was installed, whilst the original platform was lengthened so that both platforms could accommodate ten coach train formations. Although the line was double track through the platforms, both lines were signalled for bi-directional working which was justified and indeed necessary almost every Saturday for about twelve years. There was one short siding off the Down loop.

The station is 28 miles 48 chains from Barmouth and running time was seventy eight minutes. There were no freight facilities. There was a signal cabin on the Up side of the line, at the Afon Wen end, which contained a set of 27 levers. Station buildings were of simple block construction with limited enclosed toilet facilities in the centre of open fronted waiting sheds. A bridge spans the tracks at the Pwllheli end.

The double line between Penychain and Afon Wen reverted to single line after the station at Afon Wen closed.

ABERERCH

Abererch is 30 miles 15 chains and eighty two minutes running time from Barmouth. In Cambrian Railways days it was listed as a 'conditional stop' and the halt was unstaffed although there was a crossing keeper living in attendance who opened the gates for road traffic as required. The GWR upgraded the Halt to station status in 1933 with staff on duty. It remained in this capacity until 1956 when it was again reduced to an unstaffed halt. There was one siding on the Up side at the Barmouth end, access facing Pwllheli. The road crossing divided the siding from the station and the siding was unlocked by the train staff when access was required, the guard performing the duty. There was a passenger operated double arm signal to halt trains in Cambrian days. The platforms were low and steps were provided to assist passengers.

Apart from the crossing keeper's cottage which had sheet slate cladding, the station buildings were restricted to a wooden hut which contained an enclosed waiting room with a small office at the Pwllheli end. A bench seat, the obligatory name-board and lamp completed the inventory. A small 3 lever frame was located on the Down side of the track adjoining the level crossing, the levers controlling the gate lock and signals protecting the crossing. When the station was reduced in status to an un-staffed halt, the hut was removed.

169. Abererch, c. 1950. Taken from the level crossing and looking towards Afon Wen, this view shows the crossing keeper's house extreme left. Beyond that is the waiting hut, which was removed after downgrading to an unstaffed halt. The small stone platform is of low height, and a set of steps can be seen beyond the seat. The track continues on the level for about a mile towards Penychain. The station is decorated with shrubs and small flower beds, and bordered by white painted bricks. *Photo: J. H. Moss*

170. Abererch, 15th July 1965. Abererch Halt, looking towards Pwllheli, a mile and a half away. The waiting hut and the siding which was on the Up side of the line beyond the level crossing, has gone but otherwise there is little change. On the left hand side of the track can be seen the hut which housed the three lever frame, a fixed distant signal standing alongside. The once attractive flower beds have grown wild, and present a depressing picture. *Photo: C. L. Caddy*

171. Pwllheli, c. 1963. British Railways standard class 3 2-6-2T No. 82031 with a two coach working from Barmouth runs past the site of the original Pwllheli station. In the distance can be seen the balloon water tower at the shed whilst the goods stock stands alongside the goods shed at East Yard. *Photo: E. N. Kneale*

172. Pwllheli, c. 1950. Taken from near Pwllheli East box, this view shows the second single road locomotive shed at Pwllheli, with ex Cambrian Railway 0-6-0 No. 894 shunting the goods yard. The double track main line looking towards the terminus runs from the right of the picture with Pwllheli Goods Level Crossing gates across the tracks. Notice also the pre-cast concrete signal posts with the fixed distant facing the photographer. *Photo: Authors Collection*

PWLLHELI

Pwllheli is the terminus of the line, 57 miles 54 chains from Machynlleth and 32 miles from Barmouth. The original 1867 station was fifty chains closer to Afon Wen and on the extreme eastern edge of the town but with the construction of the inner harbour however, land became available and a new station was built in the centre of the town on its present site: the old station was dismantled with the platform canopy first going to Aberdovey and then, after a useful life of seventy years, on to Llanuwchllyn for the Bala Lake Railway.

The single line approached Pwllheli from Penychain and encountered Pwllheli East Frame on the Up side where the double line commenced. This controlled access to sidings on both sides of the line and also housed an intermediate key token which enabled traincrews to obtain a staff token for the single line section without having to walk back to Pwllheli East signal box located on the Up side, and housed a frame with a set of 23 levers. Access to Pwllheli locomotive shed and goods yard was at this point. From Pwllheli East Frame to Pwllheli West signalbox and the stops at the terminus was worked by Absolute Block. There was a level crossing known as 'Goods Level Crossing' between the East and West signal boxes, with a gatekeeper provided. The crossing was protected by signals with the gates normally across the tracks.

Pwllheli West signal box was located at the station throat and housed a set of forty levers.

There was an island platform with two faces at Pwllheli, covered by an attractive canopy constructed by the GWR. There were release roads alongside the platform roads and off the Up side loop line was a short siding to a loading dock. There was also storage space for coaching stock alongside the Down side loop although most coaching stock was stored at Pwllheli East.

The station buildings are of wood construction and set at right angles to the platform with the passenger access to the main Square in Pwllheli.

The locomotive shed was originally a single road wooden structure of somewhat precarious condition. It was replaced in 1959 by a new two road steel framed building with glazing running the length of the shed on both side walls. The offices were attached to one wall and the structure was 100 feet long by 45 feet wide. It was the third shed at Pwllheli and only lasted seven years although the building still stands. The shed yard included a coaling plant and a turntable which was purchased by the West Somerset Railway after closure. Locomotive allocation at Pwllheli was never listed separately, unlike Portmadoc, but was included in the Machynlleth allocation. On January 4th 1965 there were seventeen drivers and seventeen firemen shown on the roster sheet. Two weeks later, with the introduction of dmu workings in place of most of the steam duties, the arrangement structure was modified to sixteen drivers duties on the main link, with eight firemen, and a list headed 'Redundant Fireman' with nine names attached. For the record, the pairings of driver and fireman for the 4th January are shown below:

Driver *E. Jones,* **Fireman** *G. Hutchinson;* **Driver** *W. Thomas,* **Fireman** *R. Roberts;* **Driver** *E. Roberts,* **Fireman** *R. M. Morris;* **Driver** *H. M. Jones,* **Fireman** *J. A Lewis;* **Driver** *R. M. Jones,* **Fireman** *W. Samsome;* **Driver** *R. O. Jones,* **Fireman** *M. Parry;* **Driver** *J. M. Jones,* **Fireman** *R. E. Roberts;* **Driver** *J. G. Jones,* **Fireman** *P. B. Jones;* **Driver** *G. Daniels,* **Fireman** *B. Wakeley;* **Driver** *E. E. Griffiths,* **Fireman** *D. M. Williams;* **Driver** *H. M. Dobson,* **Fireman** *E. O. Roberts;* **Driver** *R. P. Evans,* **Fireman** *W. Kitchen;* **Driver** *J. G. Thomas,* **Fireman** *H. O. Evans;* **Driver** *E. E. Roberts,* **Fireman** *G. W. Jones;* **Driver** *J. J. Griffin,* **Fireman** *E. Morris;* **Driver** *R. Thomas,* **Fireman** *P. Evans;* **Driver** *W. J. Anderson,* **Fireman** *I. Roberts.*

173. (Above) Pwllheli, June 1965. The new locomotive shed was a far cry from the gloomy wooden structure that it superceded. Roomy and well lit with windows running the entire length of both shed walls, it provided comfortable working conditions for maintenance staff and traincrews engaged on preparing their locomotives. On the far wall were the shed offices, a vast improvement here also. The building still stands today but serving a very different purpose. *Photo: W. G. Rear*

174. (below) Pwllheli, June 1965. The new shed at Pwllheli called for a redesigned locomotive yard. This was the result. The balloon water tower stands at the shed disc with access to the shed roads leading off the turntable. In the background can be seen Pwllheli East signal box. The large water tank to the left hand side of the picture is typical GWR design. Pwllheli East signal cabin controlled not only the access to the goods yard and locomotive depot but also the commencement of the double track section to Pwllheli station. It was built in 1935 to replace a Cambrian Railways structure. The box housed a 23 lever frame and was finally taken out of use in 1976 and converted to a ground frame. *Photo: W. G. Rear*

175. Pwllheli, 8th September 1962. The crews of these two Collett 0-6-0 locomotives, Nos. 2257 and 2286 are involved in shunting their stock after bringing in a six coach train from Barmouth. Once the coaches were stabled the locomotives would travel the half mile to the depot for coal, water and servicing prior to taking up their next turn of duty.

Photo: Authors Collection

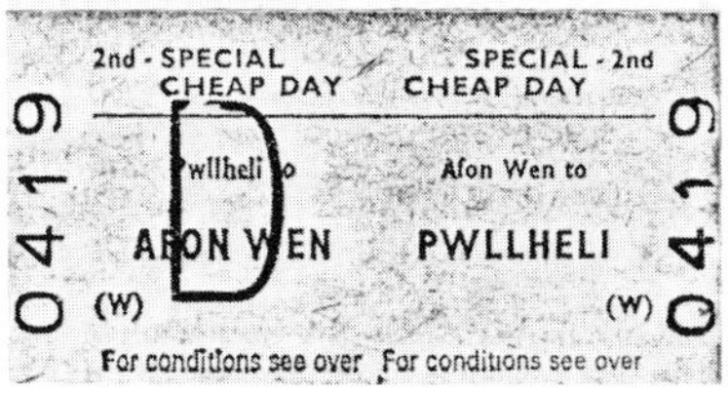

176. Pwllheli, 23rd June 1968. From the end of the platform ramp, looking east towards Afon Wen, this view shows the approach to the station and Pwllheli West signal box. The double track 'main line' are the two lines to the left of the camera whilst the lines in front of the box leads to a head shunt. Pwllheli West signal box which controlled the station area movements was of Cambrian origin and housed a frame with a set of 40 levers. To the right of the box is the inner harbour, which when constructed enabled the extension to the present station at Pwllheli to be built.

Photo: C. L. Caddy

177. Pwllheli, c. 1960. Pwllheli station comprised of a single island platform, and both faces were used for both arrivals and departures. This view, looking towards Afon Wen, shows the general layout with the tank on the end of the platform serving both lines. All signals are of lower quadrant design, and in the distance, beyond the signalbox can be seen the new locomotive shed roof.

Photo: Lens of Sutton

178. Pwllheli, c. 1960. Taken at the same time as the previous photograph, this view is taken from the end of the platform and looking towards the buffer-stops and station buildings. The four coach set for Bangor stands clear of the stops, the locomotive having run around via the loop line. Today, the right hand platform lines are no more, the site having been developed as a supermarket.

179. Pwllheli, c. 1960. Access to the platforms from the station buildings was through ticket barriers. This shows the length of the platform, its broad width and the magnificent canopy giving protection to the passengers, alas now shortened. Even with this protection, the station was exposed to the weather blowing off the Irish Sea and could be a bleak location to wait for trains. A four coach set of coaches waits for a locomotive whilst on the platform a trolley loaded with mail bags stands unguarded. *Photo: Lens of Sutton*

180. Pwllheli, 6th October 1972. An external view of the 1909 station building. *Photo: C. L. Caddy*

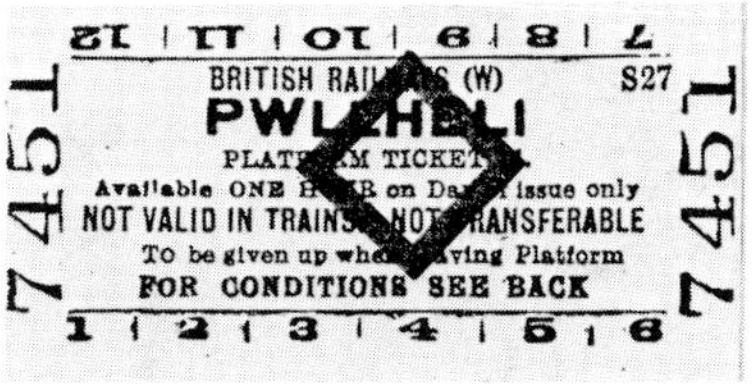

181. Pwllheli, June 1953. Perhaps not the end, but the beginning of the end!

The Stationmaster poses with some of his staff, together with Driver Moi Edwards of Bangor shed, and the guard, after working the 4.04 p.m. from Bangor. The locomotive was Bangor shed's own 41223 still bearing the wording 'British Railways'. The non corridor stock was normally found on the Amlwch branch, as was the 2-6-2T. On Saturdays, Turn 131 involved working the 12.40 p.m. Bangor to Amlwch and return, followed by the 4.04 p.m. to Afon Wen and Pwllheli, returning with the 7.05 p.m. from Afon Wen. The class 2 tanks however were masters of the job and levelled out the gradients without difficulty. On this particular day, a minor mishap occurred at Afon Wen on the return journey when the fire was cleaned at Afon Wen and the grate was 'rocked' to break up clinker forming on the bars. A segment of the grate became dislodged and eventually dropped unnoticed in the ashpan somewhere after Caernarvon. It was only when the locomotive was back at Bangor that the incident was discovered for there had been no lack of steam for the remainder of the shift. Some engine! *Photo: W. G. Rear*